Fear of Forgetfulness

Sherri Hayden

Fear of Forgetfulness

The Complexities of Cognition, Mental Health & Other Confounds

Springer

Sherri Hayden, Ph.D.
Department of Medicine, Division of Neurology
University of British Columbia Hospital
Vancouver, BC, Canada

ISBN 978-3-032-21817-9 ISBN 978-3-032-21818-6 (eBook)
https://doi.org/10.1007/978-3-032-21818-6

This Springer imprint is published by the registered company Springer Nature Switzerland AG
The registered company address is: Gewerbestrasse 11, 6330 Cham, Switzerland

For Larry Aavisto, Liam Aavisto, and Logan Aavisto.

My love and deepest gratitude for your presence and support in my life and my work.

Prologue

Some of the most challenging frontiers in clinical care are within the scope of neuropsychology, given barriers to understanding the depth and complexity of our brains. Diagnostics for many neurological conditions, including dementia, may be confounded by the complex issues pertaining to brain function. Such confounds may include but are not exclusive to mental health conditions, endocrine disorders along with co-existing psychosocial or environmental factors. This book will focus on the complex crossroads in the diagnosis of dementias within the confines of such confounds, particularly the role of mental health factors in the understanding and diagnosis of various neurodegenerative conditions.

In general terms, diagnosis of neurodegenerative disorders becomes exponentially more complex in the presence of co-existing mental health issues, psychosocial factors, and neurological and/or medical conditions. Drawing from over 30 years in neuropsychology practice, my hope is to provide a practical and clinically relevant review of such complexities in the diagnosis of neurodegenerative disorders. Such complexities have certainly been prominent in my own practice, and I am encouraged to see this acknowledgment emerging within scientific literature in recent years (Ball et al. 2020).

Over the years of my clinical experience, I have witnessed an evolution of presentation at the Clinic for Alzheimer's Disease and Related Disorders (CARD) at the University of British Columbia Hospital, where I serve as the neuropsychologist. When I first began working in this clinic in 1993, the average age of our patient population was 10 years older than it is currently. This downward shift in age of presentation cannot be solely attributable to improved detection of early onset dementias. It seems likely this change in age of patients I see in my practice may also be consequent to a combination of increased public knowledge of dementia risk as well as generational differences in health care advocacy. Along with this evolution in average age of our patient population, the numbers of patients I see with primary and co-existing mental health issues has increased substantially over my years of practice as well. We currently estimate approximately 60 percent or more of patients seen for neuropsychological assessment in this clinic present with primary or secondary mental health issues.

Such shifts in the patient populations seen in the clinic has resulted in significant changes to my approach to clinical care as a neuropsychologist, including the administration of comprehensive assessment of mood and personality traits along with the cognitive evaluation typically associated with neuropsychological assessment for neurodegenerative conditions. This form of in-depth evaluation of psychological factors is often excluded from the cognitive-focused assessments conducted within dementia clinics. In a changing clinical landscape, diagnostics in this field are challenged without such comprehensive exploration of mental health factors for those presenting to dementia investigations. The intention of this book is to provide an overview of the current knowledge regarding such factors with discussion of the important context of these and other issues complicating the presentation of neurodegenerative disorders.

Despite challenges, diagnostics in dementia has certainly been advanced by the exponential use of biomarkers and genetic factors in clinical care. However, diagnostics have also been further complicated by their use, given that such issues cannot provide a full scope of understanding for the complexity of clinical presentation in practice. This book will discuss the challenges and risks of biomarkers and genetic testing in clinical care which may lead to misinterpretation of symptom presentation, and consequently effective symptom management. The need for ongoing review of the ethical complexities for communicating biomarker and genetic information to preclinical and clinical dementia populations remains challenging (Robillard et al. 2024) and may be exacerbating mental health confounds in presentation in this population. This book will provide review of such issues in clinical practice.

Finally, the growing literature on prevention in the field of dementia has given clinicians like myself along with the patient population I serve hope for the future in this area. I often tell patients that one of the few positives from the year of 2020 when the COVID-19 struck the world was a publication by the Lancet commission on dementia prevention in which 40% of worldwide dementia cases were found to be likely attributable to modifiable factors (Livingston et al. 2020). This raises considerable hope for the impact of prevention on dementia cases worldwide. Thus, the final chapters of this book will provide a review of both prevention research and clinical compensatory strategies focused on cognition and behavior in dementia and its prodromes.

Overall, I write this book as a clinician in hopes to better support the care of those who suffer from "fear of forgetfulness." Research suggest anxiety or fear of forgetfulness and dementia may be as high as 80% worldwide (Kessler et al. 2012; Meyer et al. 2024). This book may be used by clinicians to support the care of the patients they serve through psychoeducation regarding the variety of factors that may impact both diagnostics and care in dementia. Further, for those reading this book who themselves are experiencing fear of forgetfulness, or are fearful for someone they love, my hope is to provide information that will empower you to pursue the most comprehensive diagnostics and care available in the field.

References

Ball HA, McWhirter L, Ballard C, Bhome R, Blackburn DJ, Edwards MJ, et al. Functional cognitive disorder: dementia's blind spot. Brain. 2020;143(10):2895–903.
Kessler EM, Bowen CE, Baer M, Froelich L, Wahl HW. Dementia worry: a psychological examination of an unexplored phenomenon. Eur J Ageing. 2012;9(4):275–84.
Livingston G, Huntley J, Sommerlad A, Ames D, Ballard C, Banerjee S, et al. Dementia prevention, intervention, and care: 2020 report of the lancet commission. Lancet. 2020;396(10248):413–46.
Meyer JL, Buczak-Stec E, König HH, Hajek A. Fear of dementia among middle-aged and older adults in Germany. Dement Geriatr Cogn Disord Extra. 2024;14(1):96–105.
Robillard JM, Masellis M, Martin SE, Khachaturian AS, Dixon RA. The return of biomarker results in research: balancing complexity, precision, and ethical responsibility. J Alzheimers Dis. 2024;97(3):1083–90.

Department of Medicine, Division of Neurology, University of British Columbia, Vancouver, BC, Canada
Sherri Hayden, Ph.D.

Acknowledgments

I thank my neuropsychological technician, Anne Lyle, who has provided me support and quality testing within my clinical practice over many years. I am truly grateful for your hard work and friendship.

I also would like to express my gratitude to the patients I have served over my 30+ year career, many of whom have inspired me through their "master class" on facing advancing age with dignity and grace.

Contents

About the Author

Dr. Sherri, Hayden, Ph.D., R.Psych. is a Clinical Neuropsychologist and Clinical Assistant Professor with the Division of Neurology, Department of Medicine at the University of British Columbia. She is also a faculty member of the University of British Columbia Women's Health Research Cluster.

Since 1993, she has worked at UBC Hospital in the Clinic for Alzheimer's Disease & Related Disorders as well as in her private practice at the NeuroHealth Clinic. Her clinical practices serve a variety of populations including neurodegenerative disorders, traumatic brain injuries, post-chemotherapy cognitive issues, post-COVID cognitive issues, and mental health related cognitive issues. Dr. Hayden applies preventative and lifestyle medicine principles to her clinical practice.

Part I

Progress and Challenges in Dementia Diagnostics

Prodromes of Dementia: Mild Cognitive Impairment, Biomarkers, and Genetics

1

Preclinical and Prodromes of Dementia

Perhaps one of the great challenges when navigating the vast literature within the field is understanding the heterogeneity of nomenclature pertaining to the period prior to meeting diagnostic criteria for dementia. Specifically, the terminology related to mild cognitive impairment (MCI) is complicated by the existence of varying diagnostic criteria with terminology often used interchangeably despite variance in their definitions. This heterogeneity in nomenclature has led to significant variance within the findings of research and clinical practice (Casagrande et al. 2022) in the area. Such variance serves to hinder progress toward better prediction in dementia and perhaps toward effective treatments in the future.

Thus, there remains no single, universally agreed-upon set of diagnostic criteria for mild cognitive impairment (MCI). Multiple professional bodies, such as the National Institute on Aging-Alzheimer's Association (NIA-AA) and the American Psychiatric Association (APA), have developed their own guidelines, which vary in their specific requirements (e.g., cut-off scores on cognitive tests, the number of impaired domains, the need for subjective complaints) (Sperling et al. 2011; Wilson et al. 2011; Vermunt et al. 2019). Further, several terms have been used historically and sometimes interchangeably to describe a similar clinical state regarding the prodromes of dementia (Roberts and Knopman 2013). Even when using the same set of criteria for MCI, clinicians and researchers may operationalize them differently, through using different assessment tools, cut-off values, or in the consideration of patient's education level. Such issues result in the highly variable prevalence estimates (i.e., ranging from 1.2% to 87% in one review) of mild cognitive impairment (MCI) reported in metanalysis across studies (Song et al. 2023), although global prevalence was estimated at 19.7% across the 233 studies. Notably, these researchers also found a trend within this metanalysis reflecting an increase in prevalence rates worldwide following 2019, for which they raised the need to further explore connections of MCI and COVID-19. Regardless, to illustrate this

S. Hayden, *Fear of Forgetfulness*, https://doi.org/10.1007/978-3-032-21818-6_1

Table 1.1 Terminology of prodromes in dementia

Prodome label	Description	Prevalence rates
Not cognitively impaired (NCI)	There is no objective evidence of cognitive impairment on formalized testing of cognition. Typically individuals in this category are experiencing normal age-related changes to their cognition.	25.5%–75% (Hale et al. 2020)
Subjective cognitive decline (SCI)	This refers to individuals with perceived/subjective cognitive deficits and/or impaired functional abilities. These impairments are not measurable on formalized testing and do not appear to significantly impact function (i.e., such as reported by collateral information). SCI may characterize normal aging in individuals and/or individuals with mental health issues that impact perception of cognitive performance. However, this category may also include those who are experiencing progressive decline due to a cognitive disorder, although deficits are not yet measurable on formalized assessment tools.	5%–69.1% (Röhr et al. 2020)
Preclinical Alzheimer's disease	Primarily this is identified pre-symptomatically and usually with diagnosis suggestive through use of biomarker testing and/or genetic testing. Of note, recent publications propose using biomarkers for diagnostics within clinical care settings (Jack et al. 2024).	8%–46.4% (Parnetti et al. 2019)
Mild cognitive impairment (MCI) Amnestic MCI Non-amnestic MCI Single domain MCI Multiple domain MCI	MCI entails measurable cognitive impairment present on formalized cognitive testing, although these measurable cognitive deficits are insufficient to meet criteria for dementia often without significant functional impact. Subcategories refer to patterns of cognitive impairment present on formalized testing.	3%–53.8% (Casagrande et al. 2022)

variance in nomenclature and prevalence rates, one may have seen any of the following terms used for prodromal states of dementia and/or Alzheimer's disease (Table 1.1):

Although all of these labels for dementia prodromes generally refer to the period where a patient does not meet diagnostic criteria for any variant of dementia, one can see the overlap in either perceived or measurable cognitive deficits for patients presenting within clinical settings for diagnosis of neurodegenerative disorders. This overlap contributes to diagnostic complexities within the field.

Notably, there is some disparity within the predicted conversion rates for dementia from MCI which is estimated to be approximately 10%–30% after 1 year depending upon various characteristics such as clinical setting, population type, and diagnostic methods used (Thaipisuttikul et al. 2022). More specifically, conversion rates for MCI to dementia are higher in memory clinics with rates falling between 10.2% and 33.6% over 1 year relative to community-based samples, with rates falling between 5.4% and 11.5% per year. Notably higher rates of conversion are evident when patients are followed over 3 years, with estimates upward of 42.5% within clinical settings. Despite these statistics, it is important to also acknowledge some individuals with MCI do not progress onto any form of dementia and in fact approximately 15%–30% of the MCI group appear to revert to normal cognition (Hu et al. 2017). Further complicating the issue in clinical settings is emerging evidence of increased prevalence of MCI and cognitive decline following 2019 and the COVID pandemic. In review of available literature, researchers (Song et al. 2023) have found the total prevalence of MCI in hospital settings had increased 61.7% from MCI rates prior to 2019. Such indicators support ongoing evidence for the impact of COVID-19 on cognitive function and possibly for COVID-19 related cognitive decline (Li et al. 2023). There is research to suggest conversion rate for MCI to dementia to be 55.6% for patients who contracted COVID-19 compared to a 23% rate for unaffected individuals following 18 months (Owens et al. 2023). Further, after adjustment for age, COVID-19 was associated with 4.39 times increased odds of progression from MCI to dementia. Reasons for this accelerated decline continue to be explored, although it is believed to be related to virus-induced neuroinflammation which may cause direct damage to brain cells triggering or accelerating neurodegenerative processes. Other contributing factors for these shifting rates of MCI may relate to the estimated 25% increase in mental health issues following the pandemic, which in turn may result in cognitive issues as will be discussed in subsequent chapters. Such increases in prevalence rates for MCI and conversion following 2019 has further challenged healthcare settings and providers when attempting to navigate patients' fear of cognitive decline.

Regardless of etiology, the vast variability in reported prodrome conversion prevalence rates has had direct consequences within clinical practice, including my own. This was exemplified with a 58-year-old patient who worked in academia at the time he presented to our clinic (i.e., CARD) with concerns for cognitive change. For our purposes let us call this patient George, who stated in my clinical interview he was "terrified" by the 85% conversion rate from mild cognitive impairment (MCI) to dementia that he had discovered while doing his own review of literature online prior to our meeting. He had previously been seen by a neurologist in the clinic at which time he was directly informed of the MCI label. Following this neurology consultation, which occurred approximately 3 months prior to my meeting with him, George decided to do his own research online. He did express feeling the MCI label had not been sufficiently explained to him during his neurology consult, although it was unclear if this was indeed the case and to what extent his perception of the process impacted this view. Nevertheless, within my neuropsychological evaluation, George was found to have considerable mental health factors (i.e., marked anxiety at approximately 98% on psychological measurement) that were likely contributing to the modest and variable cognitive deficits present on neuropsychological evaluation. Unfortunately, such mental health factors were not considered or discussed when the label of MCI was provided to George. Despite George's relatively sophisticated knowledge of research methodology, his marked anxiety superseded this understanding of the variance reflected within research.

Overall, this lack of a clear, standardized diagnostic methodology makes it difficult to compare results across studies; hinders the development of effective, targeted treatments; and can lead to confusion for patients and increased clinical complexity for healthcare providers.

With the exception of the prodromes where labeling relies on biomarker results, which will be discussed in more detail later, prodromal labels are largely based upon a patient's performance on formalized cognitive testing coupled with clinical judgment within the context of available neuroimaging. The complexity of such labeling is further increased by mental health factors that may distort not only an individual's perception and/or interpretation of their own cognitive performances but may also impact clinical judgment regarding a patient's presentation (Stella et al. 2014). Within my one career over the years, it is this issue that has consumed much of my clinical practice and inspired the writing of this book.

Perhaps one of the most challenging issues I face within the context of clinical neuropsychology practice is navigating the nuance in educating patients regarding the issues around the prodromes of dementia and the meaning of this to them on a functional basis. An example of such challenge is well illustrated by a patient seen in my practice, we can call her Lara, who was an Asian female aged 64 years attending the Clinic with her adult daughter. Lara had been followed by one of the team neurologists for several years and had been told she was MCI based upon her performance on cognitive screening testing, the MoCA (i.e., a brief cognitive screening test which provides rudimentary evaluation of multiple cognitive domains with a total score out of 30). The MoCA is an extremely useful tool for physicians as a basic cognitive screening tool, and is typically the measurement used by physicians to identify mild cognitive impairment (Islam et al. 2023). However, within the complexity of clinical settings, there are often considerable limitations to account for low scores on cognitive screens such as the MoCA particularly when confounds of mental health factors on cognitive performance exist. Thus, longitudinal cognitive screening evaluations (i.e., serial MoCAs) and in some cases, more in-depth cognitive evaluation as is available through neuropsychological testing can provide some clarity for such clinical presentation.

So back to our patient, Lara had been concerned for her perceived cognitive issues over at least 4–5 years. Lara felt her cognition may be worsening over time, although she acknowledged this was possibly somewhat variable in her daily life. Lara came for a neuropsychological assessment to provide diagnostic opinion as her MoCA score had not changed substantively over a couple of years of being followed by our clinic neurologist, despite Lara's perception that her cognition may be worsening at times. Within the neuropsychological assessment, she performed relatively well with the exception of a borderline score within a single measure of confrontation naming. This performance was likely consequent to her history of English as a Second Language (ESL) and cultural factors (i.e., born and educated in Hong Kong). During the feedback session following her neuropsychological assessment with Lara and her daughter, we discussed the discrepancy between Lara's perception of cognitive function and the measured performance on neuropsychological testing. Lara's daughter had indicated within interview for this assessment that her mother was averse to acknowledging mental health factors, which was affirmed within my separate interview with Lara who stated she had “no mood issues” although she did acknowledge having been a “worrier” in the past. As mentioned previously, within the context of neuropsychological assessment, I include comprehensive assessment of mental health factors (i.e., mood and personality factors) with embedded validity scales. Within this measurement of mental health, Lara’s responses reflected a degree of positive impression management, likely resulting in minimization of psychological symptoms possibly related to repression and/or denial of such symptoms. Despite this context of psychological symptom minimization, Lara’s responses revealed marked anxiety symptoms, consistent with need to rule out a generalized anxiety disorder (GAD). A lengthy discussion was required to provide Lara with context for this anxiety as she described feeling no overt signs of anxiety, again despite acknowledging “worry.” It was clear that Lara did not connect feelings of “worry” with any form of anxiety. Regardless, our lengthy discussion did result in Lara eventually acknowledging the possibility of anxiety for which she was agreeable to the recommended treatment of psychotherapy. Without this formalized assessment of mood and personality traits as well as more in-depth testing of cognitive function within neuropsychological assessment, Lara would not have come to acknowledge the role of mental health within her clinical presentation.

Certainly, Lara’s case emphasizes the critical need for persual of treatment for mental health factors within the prodromes of dementia, particularly given emerging evidence within scientific literature of such mental health issues as risk factors for cognitive decline (Mirza et al. 2017). Such research suggests the high rates of depression and anxiety in MCI do not just represent symptom presentation but may speed progression to dementia in some individuals and certainly may impact memory on a variable basis in many, even if their cognitive function does not decline over time. Overall, it is clear that managing such mental health conditions is essential as they may represent reversible causes of cognitive issues impacting lifestyle.

Biomarkers and Genetics

Biomarker and genetic testing are emerging as an exciting means of mitigating some of the aforementioned complexities present within diagnosis of neurodegenerative disorders and their prodromes. However, with the advent of biomarker and genetic testing come a different set of challenges within the clinical frontlines, particularly with respect to communications about limits of such testing to patients presenting with fear of cognitive decline. In our own clinic, genetic counselors provide such communications regarding genetic testing directly to patients and their families, while physicians provide communications regarding biomarker testing.

Prior to discussing such challenges with use of such testing in some clinical cases, a brief review of biomarkers and genetics of prodromal dementia is warranted. Notably, this review of biomarkers and genetics in dementia will be rudimentary in nature and limited in scope with the intention of simply providing a foundation from which discussion of the implications for clinical care and management can arise.

Biomarkers of Prodromal Dementia

Biomarkers, in very simple terms, represent the "indicator lights" which may reflect underlying pathological changes associated with dementia. These biomarkers are measurable and typically found within testing of body fluids, including as blood and spinal fluid, as well as neuroimaging such as PET scans and MRI scans of the brain (AlMansoori et al. 2024). Biomarkers may be generally categorized into those that indicate core pathological changes which are typically associated with Alzheimer's disease. These pathological changes (i.e., the physical or functional changes to body cells, tissues, or organs that are caused by disease) are hypothesized to begin 10–20 years *before* individuals become symptomatic for the disease (de Vidania et al. 2020; Zhang et. al. 2025) So, these core pathological changes of Alzheimer's disease are reflected in the following:

- *Amyloid-beta Plaques*—These are the insoluble clumps of "trash" or plaques which build up outside of brain cells although in normal brains are cleared away. In Alzheimer's disease, these plaques accumulate. Biomarker testing for amyloid-beta plaques may be measured using Amyloid PET scans to visualize plaques directly or via cerebrospinal fluid (CSF) testing which can show a drop in amyloid-42 levels reflecting the fact that it has not being cleared.
- *Tau Tangles*—It is a protein normally required for the stabilization of the brain cells' transportation system. In dementia, tau breaks down this transportation system due to hyperphosphorylation (i.e., becomes overloaded with phosphate groups leading to tangles) and causing cell death. Biomarker testing for Tau Tangles includes Tau PET scans to show the spread of tangles or via CSF tests to reveal high levels of phosphorylated tau (p-tau).

The other category of biomarkers are those that reflect downstream neurodegeneration, in other words, the physical damage caused by the core pathology. These downstream indicator lights are most directly associated with symptoms seen within clinical setting (Grande et al. 2025). Such downstream indicator lights are reflected in the damage, death, and shrinking of neurons caused by amyloid and tau. Such downstream changes may be tested through such investigations as MRI scans (i.e., revealing atrophy or shrinkage particularly in the hippocampus) or FDG PET scans (i.e., reflecting lowered glucose metabolism which suggests brain cell death or dysfunction).

Despite these scientific advances with the development of biomarkers for diagnostics in dementia, there remain challenges for their clinical use. Such issues are inherent within the context of diagnostic uncertainty of dementia prodromes described earlier. It is also difficult to accurately differentiate between MCI that will progress to Alzheimer's disease (AD) versus MCI that evolves into other types of dementia (e.g., dementia with Lewy bodies, vascular dementia), or MCI that may remain stable or even revert to normal function. Conflicting results across different biomarker modalities (e.g., amyloid beta and Tau measures) at the individual level are common and difficult to interpret within the diagnostic uncertainty of the dementia prodromes. Further, there can be a lack of specificity and predictive power as many biomarkers are specific to AD but may be altered in other neurodegenerative disorders. Overall, it is challenging to differentiate individuals whose MCI will progress to dementia from those who will remain stable using current markers, particularly in the preclinical stage.

Further confounds to biomarker testing include the use of different methods and techniques for biomarker measurement and data normalization across studies which then makes comparing data reported by different groups difficult, impeding progress in the field. Also, biomarker results can be influenced by pre-analytical handling of target materials as well as patient comorbidities (i.e., including mental health and coexisting health conditions such as hypertension or kidney disease), and demographic factors (i.e., age, race, and ethnicity) that have not been well represented in validation studies. All of these confounding factors emphasize the need for caution in clinical interpretation of biomarkers across diverse populations. Finally, for cognitively unimpaired individuals (i.e., not cognitively impaired—NCI), there is currently no evidence that biomarker testing improves outcomes within the context of lack of effective disease-modifying treatments for many neurodegenerative disorders once diagnosed leading to complications for the decision to test using biomarkers.

Generally, given these current challenges to clinical use of biomarker testing in diagnosis of dementia, it is recommended that neuropsychological testing remain an important component within the diagnostic process for dementia (Carlos and Josephs 2023). This is particularly pertinent, as studies have shown some individuals may have the core pathological biomarkers of disease (e.g., amyloid plaques) but exhibit normal cognitive function (de Vries et al. 2024). Relying solely on a positive biomarker in such a case could lead to overdiagnosis, causing unnecessary psychological burden in such cases. Further, a mismatch can occur between a patient's biomarker levels and their subjective experience or overall sense of well-being. Biomarkers may track a disease process but not necessarily the actual functional decline experienced by the patient, which is the primary concern in clinical care and management. I have certainly seen this exemplified within clinical practice as I typically will provide neuropsychological testing months or in some cases years, following neurology consultation where biomarker testing may have been discussed with the patient and their families. I am seeing increasing numbers of patients and their families in considerable distress, often including anticipatory anxiety and/or catastrophic reactions to being informed of their results. I will discuss issues associated with communications regarding biomarkers later in this chapter. Regardless, overall, biomarkers may provide invaluable insight into the underlying biological changes within dementia diagnosis and care, although they cannot fully substitute a patient-centered clinical assessment that reflects how a patient feels and functions as is provided within comprehensive clinical and neuropsychological evaluations.

Genetics of Prodromal Dementia

In addition to biomarkers, genetic factors play a role in determining the risk of developing dementia in some cases along with the rate of progression from prodromal states, such as MCI. It is important to note, and I often indicate to my own practice, that although genetic forms of dementia are often of primary concern to patients it is believed to account to a very small portion, approximately 1%–5% of the patients that we see.

To simplify a complex area, one might view genetic testing in dementia as a means for analyzing individuals' DNA to determine change (i.e., mutations) or variations related to these diseases (Lourida et al. 2019). Regardless, there are two types of genetic testing when deemed appropriate to pursue:

- *Diagnostic testing*—This testing is conducted on those affected by the disease, often under age 65. The purpose of this testing is to determine the presence of a faulty gene and can confirm diagnosis in affected patients.
- *Predictive testing*—This testing is conducted in family members of individuals known to carry a causal gene to determine if they carry the same gene. This can obviously be life-altering information and must be communicated in a process typically managed by genetic counselors.

Furthermore, there are two different categories of genes to be tested within the above context. These include the following:

- *Deterministic Genes*—These genes directly cause a specific disease and if the gene is inherited, the individual is guaranteed to develop the disease. Notably, deterministic genes are relatively rare but include early onset familial forms of Alzheimer's disease (PSEN1, PSEN2).
- *Susceptibility Genes*—These genes reflect increased risk or predisposition but do *not* guarantee disease onset. This includes the APOE-4 allele that is the most widely recognized genetic risk factor for late onset Alzheimer's disease although this is not a definitive diagnostic tool due to variable sensitivity and specificity.

As with biomarker testing, genetic testing comes with multiple challenges to clinical care (Porteri and Frisoni 2014). Of note, late-life cognitive impairment, as would be suggested with APOE-4 risk, is complex being influenced by both genetic and environmental factors. Currently, known genetic variants explain only a small proportion of the total genetic contribution of various forms of dementia. Also of significance, genetic testing often cannot determine if an individual will become symptomatic or how severe symptoms will become or in what timeline of progression of symptoms over time.

Given these complexities in understanding etiology and predictability of dementia prodromes, one would assume that the more recent developments pertaining to biomarkers and genetic testing would simplify the clinical process. There is no doubt that this form of testing has advanced diagnostic care to date. This area is evolving with recently revised biologically diagnostic criteria for Alzheimer's disease proposed by the Alzheimer's Association Workgroup (Jack et al. 2024) to include biomarker/genetic testing for diagnostics. However, this working group of scientists do acknowledge the challenges in implementing such criteria universally. Thus, although biomarkers/genetic markers hold potential for greater diagnostic specificity in Alzheimer's disease, the complexity of effectively communicating such information to patients remains, particularly in a disease with no viable cure at present.

Although much more exploration of this area is required, emerging research suggests the need for understanding contextual factors to better inform personality- and culturally sensitive biomarker testing to ensure minimization of any negative psychological impact when informing patients of their biomarker/genetic marker risks. A 2023 study (Clark et al. 2023) focusing on these factors and their association with anticipated psychological impact, including distress, stigma, and cognitive symptoms found factors such as female gender, lower internal health locus of control (i.e., lower belief that health is under one's control), and higher concern about Alzheimer's disease are all significant predictors of anticipated distress. The intent of this book is not to focus upon the specifics of biomarker and/or genetic marker testing in dementia, but rather to discuss the issues relating to the potential psychological issues in response to provision of biomarker/genetic marker testing results in clinical settings.

Certainly, the need for understanding nuance in communications particularly in the presence of divergence in clinical presentation surrounding biomarker or genetic testing has been a challenge within my own clinical practice. This was evident in a 63-year-old patient, we will call Bob, who was informed of his positive biomarker testing (i.e., CSF testing—depressed A-beta 42 and elevated phosphorylated tau) consistent with underlying Alzheimer's pathology. However, on neuropsychological testing, Bob did not meet diagnostic criteria for dementia although his performance reflected variability in verbal memory on measurement with relatively reserved function in all other cognitive domains. Of note, Bob's medical history included a couple of concussions, with mental health history including significant panic episodes in his 20s as well as subsequent anticipatory anxiety (i.e., stating he has had worry about having additional panic episodes of that nature throughout his life since). Psychological testing within the neuropsychological assessment revealed severe anxiety-based personality traits (i.e., in the 100th percentile on anxiety screening). Within this context of mental health history, Bob was informed of his biomarker results which unfortunately triggered a catastrophic reaction including considerable anxiety and hopelessness. His partner reported he would often become tearful throughout the day and did not want to get out of bed "because of this disease." Clearly, Bob's mental health history was an important factor in predicting his psychological distress in response to results of biomarker testing, particularly prior to evidence of consistent cognitive issues on measurement.

Historically within the scope of my clinical practice, patients coming to the clinic would experience anxiety and/or distress in response to experienced cognitive change in their daily lives. Over time, such patients would often experience gradual psychological adjustment informed by cognitive performance on testing in clinical practice, which then could be supported through cognitive compensatory strategies and psychological intervention. With biomarker/genetic marker testing, the distress triggered for patients, such as Bob, most likely represents anticipatory anxiety related to the possibility of cognitive change that is to come in the future. Within this new frontier of dementia diagnostics using biomarkers/genetic testing, focusing on the empowerment of compensatory strategies proves ineffective in reducing this form of anticipatory distress for future decline. Thus, it will be critical to consider factors such as mental health history and personality factors when communicating the nuances of risk when referencing the biomarker/genetic marker testing in patients seen in clinical settings. Further, when considering such mental health factors, ensuring there are adequate psychological supports is essential when communicating biomarker results.

Overall, the prediction and labeling of the prodromes of dementia remain exceedingly complex despite the exciting scientific developments of biomarker and genetic testing. Considered clinical assessment, including neuropsychological and psychological evaluations, will remain critical for individualized clinical care, particularly in cases where there may be premorbid or coexisting mental health factors as the heterogeneity of dementia prodromes is exacerbated by such factors. The nature of such confounding and/or exacerbating factors will be discussed in greater detail in future chapters, with particular focus upon mental health factors. With greater clinical reliance upon biomarker and genetic testing within memory clinics specialized in dementia diagnostics, the risk for related iatrogenic mental health conditions may be heightened. Such risks raise the requirement for increased clinical resources aimed at addressing such mental health conditions within memory clinics, in addition to the need for ongoing review of clinical approaches to care in order to mitigate such risks for iatrogenic harm.

References

AlMansoori ME, Jemimah S, Abuhantash F, AlShehhi A. Predicting early Alzheimer's with blood biomarkers and clinical features. Sci Rep. 2024;14(1):6039.

Alzheimers.gov. What is mild cognitive impairment? n.d. https://www.alzheimers.gov/alzheimers-dementias/mild-cognitive-impairment.

Carlos AF, Josephs KA. The role of clinical assessment in the era of biomarkers. Neurotherapeutics. 2023;20(4):1001–18.

Casagrande M, Marselli G, Agostini F, Forte G, Favieri F, Guarino A. The complex burden of determining prevalence rates of mild cognitive impairment: a systematic review. Front Psych. 2022;13:960648.

Clark LR, Erickson CM, Chin NA, Basche KE, Jonaitis EM, Ketchum FB, Gleason CE. Anticipated psychological or behavioral reactions to learning Alzheimer biomarker results: associations with contextual factors. Alzheimer Dis Assoc Disord. 2023;37(4):282–9.

de Vidania S, et al. Prodromal Alzheimer's disease: constitutive upregulation of neuroglobin prevents the initiation of Alzheimer's pathology. Front Neurosci. 2020;14:562581.

de Vries LE, Huitinga I, Kessels HW, Swaab DF, Verhaagen J. The concept of resilience to Alzheimer's disease: current definitions and cellular and molecular mechanisms. Mol Neurodegener. 2024;19(1):33.

Grande G, Valletta M, Rizzuto D, et al. Blood-based biomarkers of Alzheimer's disease and incident dementia in the community. Nat Med. 2025;31:2027–35.

Hale JM, Schneider DC, Mehta NK, Myrskylä M. Cognitive impairment in the U.S.: lifetime risk, age at onset, and years impaired. SSM Popul Health. 2020;11:100577. https://doi.org/10.1016/j.ssmph.2020.100577. Erratum in: SSM Popul Health. 12:100715.

Hu C, Yu D, Sun X, Zhang M, Wang L, Qin H. The prevalence and progression of mild cognitive impairment among clinic and community populations: a systematic review and meta-analysis. Int Psychogeriatr. 2017;29(10):1595–608.

Islam N, Hashem R, Gad M, et al. Accuracy of the Montreal Cognitive Assessment tool for detecting mild cognitive impairment: a systematic review and meta-analysis. Alzheimers Dement. 2023;19:3235–43.

Jack CR, Andrews SJ, Beach TG, et al. Revised criteria for the diagnosis and staging of Alzheimer's disease. Nat Med. 2024;30:2121–4.

Li Z, Zhang Z, Zhang Z, Wang Z, Li H. Cognitive impairment after long COVID-19: current evidence and perspectives. Front Neurol. 2023;14:1239182.

Lourida I, Hannon E, Littlejohns TJ, Langa KM, Hyppönen E, Kuzma E, Llewellyn DJ. Association of lifestyle and genetic risk with incidence of dementia. JAMA. 2019;322(5):430–7.

Mirza S, Ikram MA, Bos B, Mihaescu R, Hofman A, Tiemeier H. Mild cognitive impairment and risk of depression and anxiety: a population-based study. Alzheimers Dement. 2017;13(2):130–9.

Nasreddine ZS, Phillips NA, Bédirian V, Charbonneau S, Whitehead V, Collin I, Cummings JL, Chertkow H. The Montreal Cognitive Assessment, MoCA: a brief screening tool for mild cognitive impairment. J Am Geriatr Soc. 2005;53(4):695–9. https://doi.org/10.1111/j.1532-5415.2005.53221.x. Erratum in: J Am Geriatr Soc. 2019;67(9):1991. doi:10.1111/jgs.15925. PMID: 15817019.

National Academies Committee on Preventing Dementia and Cognitive Impairment. Preventing cognitive decline and dementia: a way forward. Washington, DC: The National Academies Press; 2017. https://nap.nationalacademies.org/catalog/24782/preventing-cognitive-decline-and-dementia-a-way-forward.

Owens CD, Bonin Pinto C, Mukli P, Szarvas Z, Peterfi A, Detwiler S, Olay L, Olson AL, Li G, Galvan V, Kirkpatrick AC, Balasubramanian P, Tarantini S, Csiszar A, Ungvari Z, Prodan CI, Yabluchanskiy A. Vascular mechanisms leading to progression of mild cognitive impairment to dementia after COVID-19: protocol and methodology of a prospective longitudinal observational study. PLoS One. 2023;18(8):e0289508.

Parnetti L, Chipi E, Salvadori N, D'Andrea K, Eusebi P. Prevalence and risk of progression of preclinical Alzheimer's disease stages: a systematic review and meta-analysis. Alzheimers Res Ther. 2019;11(1):7.

Porteri C, Frisoni GB. Biomarker-based diagnosis of mild cognitive impairment due to Alzheimer's disease: how and what to tell. A kickstart to an ethical discussion. Front Aging Neurosci. 2014;6:41.

Roberts R, Knopman DS. Classification and epidemiology of MCI. Clin Geriatr Med. 2013;29(4):753–72.

Röhr S, Pabst A, Riedel-Heller SG, Jessen F, Turana Y, Handajani YS, Brayne C, Matthews FE, Stephan BCM, Lipton RB, Katz MJ, Wang C, Guerchet M, Preux PM, Mbelesso P, Ritchie K, Ancelin ML, Carrière I, Guaita A, Davin A, Vaccaro R, Kim KW, Han JW, Suh SW, Shahar S, Din NC, Vanoh D, van Boxtel M, Köhler S, Ganguli M, Jacobsen EP, Snitz BE, Anstey KJ, Cherbuin N, Kumagai S, Chen S, Narazaki K, Ng TP, Gao Q, Gwee X, Brodaty H, Kochan NA, Trollor J, Lobo A, López-Antón R, Santabárbara J, Crawford JD, Lipnicki DM, Sachdev PS, for Cohort Studies of Memory in an International Consortium (COSMIC). Estimating prevalence of subjective cognitive decline in and across international cohort studies of aging: a COSMIC study. Alzheimers Res Ther. 2020;12(1):167.

Song WX, Wu WW, Zhao YY, Xu HL, Chen GC, Jin SY, Chen J, Xian SX, Liang JH. Evidence from a meta-analysis and systematic review reveals the global prevalence of mild cognitive impairment. Front Aging Neurosci. 2023;15:1227112.

Sperling RA, Aisen PS, Beckett LA, et al. Toward defining the preclinical stages of Alzheimer's disease: recommendations from the National Institute on Aging-Alzheimer's Association workgroups on diagnostic guidelines for Alzheimer's disease. Alzheimers Dement. 2011;7:280–92.

Stella F, Radanovic M, Balthazar ML, Canineu PR, de Souza LC, Forlenza OV. Neuropsychiatric symptoms in the prodromal stages of dementia. Curr Opin Psychiatry. 2014;27(3):230–5.

Thaipisuttikul P, Jaikla K, Satthong S, Wisajun P. Rate of conversion from mild cognitive impairment to dementia in a Thai hospital-based population: a retrospective cohort. Alzheimers Dement (N Y). 2022;8(1):e12272.

Vermunt L, Sikkes SAM, van den Hout A, Handels R, Bos I, van der Flier WM, Kern S, Ousset PJ, Maruff P, Skoog I, Verhey FRJ, Freund-Levi Y, Tsolaki M, Wallin ÅK, Olde Rikkert M, Soininen H, Spiru L, Zetterberg H, Blennow K, Scheltens P, Muniz-Terrera G, Visser PJ, Alzheimer Disease Neuroimaging Initiative; AIBL Research Group; ICTUS/DSA Study Groups. Duration of preclinical, prodromal, and dementia stages of Alzheimer's disease in relation to age, sex, and APOE genotype. Alzheimers Dement. 2019;15(7):888–98.

Wilson RS, Leurgans SE, Boyle PA, Bennett DA. Cognitive decline in prodromal Alzheimer disease and mild cognitive impairment. Arch Neurol. 2011;68(3):351–6.

Zhang J, Kong G, Yang J, Pang L, Li X. Pathological mechanisms and treatment progression of Alzheimer's disease. Eur J Med Res. 2025;30(1):625.

Measuring Cognition in the Diagnosis of Dementias

2

Assessment of Cognition in Dementia Diagnostics

The role of formal cognitive assessment is critical when individuals are presenting with subjective cognitive complaints. There remains considerable debate, both in research and clinical practice, regarding the significance of subjective cognitive decline in the prediction of future cognitive decline. Researchers, such as Basagni and Narvarrete, emphasized the importance of both comprehensive neuropsychological testing and consideration of "critical factors" such as mood, personality, menopause, and other factors to be incorporated within this process (Basagni and Navarrete 2025). For this reason, to understand the presentation of cognitive issues in dementias and their prodromes, I will first review the process of cognitive assessment.

As a neuropsychologist, my primary focus in practice has been the formalized assessment of cognitive, mood, and personality factors in patients' clinical presentation for both neurological and psychiatric conditions. Within this context, neuropsychologists use standardized cognitive and mood/personality tests (i.e., paper and pencil or computerized tasks) to provide a conceptual "map" of brain function. This involves administration of the standardized cognitive tests to address diagnosis,

S. Hayden, *Fear of Forgetfulness*, https://doi.org/10.1007/978-3-032-21818-6_2

planning, and treatment for neurocognitive issues identified upon referral. Each neuropsychologist will design their own battery (i.e., selection) of tests along with determining the order in which it these tests are administered. Of note, the variability in test batteries chosen by neuropsychologists may impact comparability of findings between neuropsychologists to some degree, although typically there is some degree of overlap of tests utilized. Certainly, within diagnostics for neurodegenerative disorders where longitudinal follow-up is an important component of the investigative process, some degree of consistency in test battery is critical to ensure accurate monitoring of disease course.

Notably, within neuropsychological evaluations there are typically both *performance-based* tasks (i.e., tests that require the patient to perform a task some of which may be timed) and *self/collateral report* tasks (i.e., tests based solely on the patient or their collaterals' report of symptoms) utilized. Performance-based tasks are less susceptible to biases of perception so should be included within all comprehensive evaluations. It is important to recognize that when a patient or their loved ones' observations of cognitive/functional change is solely relied upon for clinical judgment, the role of psychological factors in the perception of cognitive performance can be underestimated or misconstrued. Nevertheless, a patient and their loved one's perception of change in function and/or cognition is an important piece to ensure a comprehensive understanding dementia and its prodromes particularly with functional cognitive disorders to be discussed in future chapters.

Ideally, neuropsychological test batteries are designed to tap into *all* measurable cognitive domains, as illustrated below. Such test batteries typically include multiple tasks within each domain to allow monitoring of consistency within domain performance. This is particularly important in cases with complex comorbidities, such as mental health factors which can create cognitive patterns mimicking dementia as will be discussed in later chapters. An "indicator light" for such mental health issues presenting cognitive change is often reflected in *convergent validity*, which essentially refers to consistency in finding of measurement for the same function or construct. As will be discussed in subsequent chapters, cognitive dysfunction primarily related to mental health factors such as functional cognitive disorders tend to have low convergent validity on neuropsychological testing (Cabreira et al. 2025).

Although tasks may be identified to measure specific cognitive performances (i.e., memory tests and language tests), neuropsychologists understand the multifactorial nature of all tests. Cognitive tests rarely measure a single aspect of cognitive function despite having a primary focus of evaluation of function. So, for example, verbal memory tests often require language skills, both expressive and receptive, to understand and respond to the memory task. It is in the purview of neuropsychologists to understand these psychometric properties of the assessments tools they chose to include within their neuropsychological test batteries. As with brain structures, cognitive domains are not discrete entities, and there is considerable interdependency in all of our cognitive abilities (Fig. 2.1).

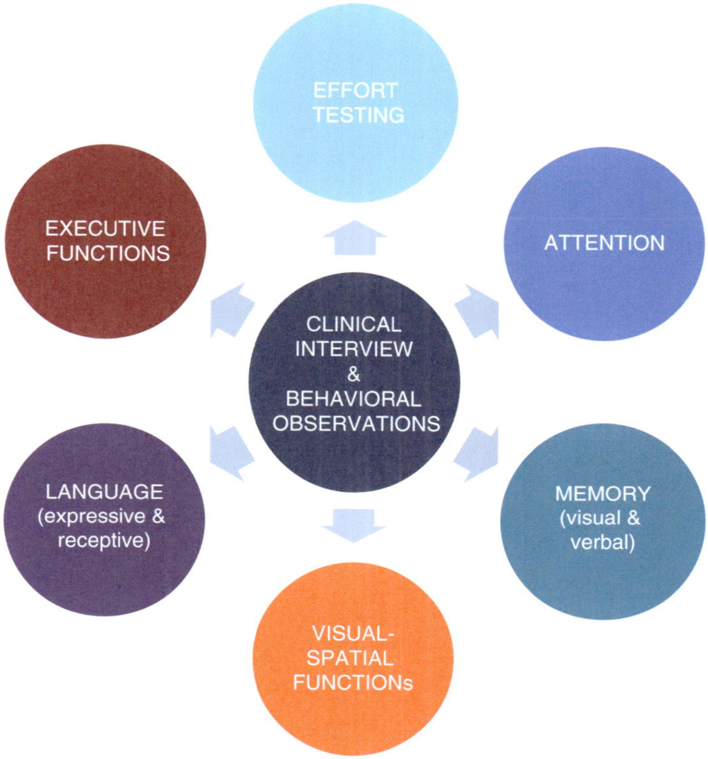

Fig. 2.1 Components of neuropsychological assessment

Cognitive Domains

With regard to procedures within neuropsychological testing, prior to administration of cognitive testing, the neuropsychologist will conduct a clinical interview, which in my own practice is semi-structured (i.e., with predetermined questions although there is opportunity to veer off of question topics if deemed clinically relevant). The clinical interview typically informs the choices made for the appropriate test battery. Clinical interview is also an excellent opportunity for the neuropsychologist to explore patient and family perceptions/biases regarding their functional status which may ultimately inform conclusions when there are discrepancies between measured and perceived cognitive performance (Table 2.1).

In my experience, change or decline within various cognitive domains perceived by patients is often associated with considerable distress and frustration as such areas of function impact all aspects of our daily lives. For example:

- Without sustained attention, it can be difficult to learn or remember information later as it was never encoded during times of distraction. This will result in the presentation of sporadic or variable memory deficits, depending upon context.

Table 2.1 Cognitive domains

Cognitive domain	Descriptor
Attention	The capacity to sustain and direct thoughts/actions within a task as well as across tasks (i.e., over time). This may also focus upon a target item while ignoring competing (irrelevant material) or to attend to more than one source of information at a time. The latter may be referred to as divided attention.
Memory (visual & verbal)	An individual's ability to acquire (learn), retrieve (recall) and retain (store) information over time. These tasks measure short-term memory, which represents a temporary store for recently presented information.
Language (expressive & receptive)	This domain involves measuring both expressive (i.e., ability to communicate/express ideas verbally) and receptive (i.e., ability to comprehend verbal information) language skills.
Visual-spatial & visual-motor skills	The capacity to understand and manipulate visually presented material as well as understand the spatial relationships of such information. Some of these tasks are timed and may be affected by motor speed as well.
Executive functions	A complex range of abilities that includes the capacity to control and organize information. This domain also involves reasoning, judgment, the ability to plan and carry out behavior, as well as to shift efficiently from one concept to another (mental flexibility). The capacity to initiate responses when needed and to inhibit inappropriate responses is also included.

- Without retention of information in our memory system, we lose the feeling of continuity in our lives making it difficult to reflect on experience to enable positive change in behavior and to plan for the future. This feeling of discontinuity, often experienced in the prodromes of dementia, can lead to increased anticipatory anxiety and maladaptive behavior patterns aimed at controlling such anxiety.
- Without preservation of language skills, we encounter difficulty communicating with others in our lives leading to feelings of isolation and/or social withdrawl. The isolation associated with language issues may also lead to increased mental health issues, such as anxiety or depression.
- Without our complex executive skills, we will struggle with screening out unimportant information in our daily lives. We may also have trouble with planning ahead or in the initiation and execution of taking actions toward the things we hope to achieve. Such issues can lead to sporadic forgetfulness, for reasons similar to difficulties in attention, due to inefficient or disorganized encoding of information. It may also lead to social withdrawal as well as increased mental health issues including depression and anxiety in response to feeling overstimulated in social environments and/or feeling overwhelmed due to disorganization. Further, in some cases, executive dysfunction results in disinhibition (i.e., lack of filter to control expression of emotion or behavior) which may create considerable social and interpersonal consequences (i.e., such as strained relationships) leading to increased risk for mental health issues.

Along with measurement of all cognitive domains, effort testing is also typically administered (i.e., stand-alone tests or embedded components of tests to determine patient's effort and consistency) to allow the neuropsychologist to determine the validity of testing being conducted for each patient. Further, behavioral observations during testing are an additional critical means of determining a patient's engagement and effort in the testing sessions. In my own practice, these behavioral observations are provided by my neuropsychological technician who conducts the cognitive testing while making detailed notes of behavior during the session. Such observations during testing are invaluable to understanding a patient's response to cognitive challenges as well as insight into approach to tasks (i.e., perfectionistic, hypervigilant, overly negative or positive self-evaluation). All of these indices can be important clues to how a patient is perceiving and/or reacting to any existing or anticipated cognitive issues occurring in their daily lives, which may be utilized as means to determine the role of mental health factors in the clinical presentation, as will be discussed within subsequent chapters.

Once testing is complete, test protocols are scored and compared to normative data (i.e., typically controlling for factors such as age and education). These normed neuropsychological test results may then be used to correlate cognitive function with neuroimaging and/or other medical findings to aide diagnostics. Comparison of results to normative data sets is interpreted by the neuropsychologist in the context of reported history, available documentation/collateral information, and behavioral indices observed during testing and interview.

Of note, most of the tests used by neuropsychologists are only accessible to licensed psychologists/neuropsychologists. This allows for protection of such tests to limit practice effects, in that by limiting public access individuals will not have prior exposure to the protocols. This ensures individuals cannot "practice" the tests which in turn would distort results and as a consequence impact diagnoses and conclusions.

An important area of expertise for neuropsychologists is the understanding of psychometric properties of the tests being used in order to incorporate such factors into the interpretive analysis. Indeed, multiple factors must be considered and balanced within the interpretation of neuropsychological test results and are illustrated below, which highlights the fact that the interpretative process for cognitive testing entails more than a simple review of test scores. The quality of this clinical interpretation can be somewhat dependent on the experience/expertise of the neuropsychologist reviewing the data, as more experience can certainly inform interpretation.

Overall, a typical neuropsychological assessment will include the following components:

- Semi-structured clinical neuropsychological diagnostic interview with patient to explore perception of cognitive performance within each domain as well as review of functional impact of perceived symptoms.
- Collection of collateral information through interview conducted separately from interview with patient to ensure open disclosure of observations albeit with patient's prior consent. Once again, this provides insight into functional impact and may highlight issues with perception/misperception of patient and/or within the kinship group.
- Four to six hours of testing (or more) administered over 1–2 days (or more), including breaks as needed.
- Evaluation within all cognitive domains as described earlier, including more than one test per domain (i.e., to determine convergent validity).

- Include standardized and comprehensive mood/personality testing.
- May be conducted in various formats:
  - Entirely in-person: This is the traditional format for neuropsychological evaluations with all aspects of cognitive and psychological testing occuring in person within the clinical setting.
  - Entirely virtually: This is an increasingly more common means of neuropsychological evaluation since the COVID-19 pandemic and entails interview and testing conducted via secured virtual platform. Research has found virtual neuropsychological assessments conducted by experienced clinicians to be a reliable form of evaluation (Brown and Zakzanis 2025). Certainly, in my own practice, this has improved access to assessment for patients in remote communities, who in the past would have had to travel great distance to access service. However, of note, I do not conduct virtual evaluations with patients who appear to have marked cognitive impairment and those who are unable or unwilling to use associated technology.
  - Hybrid of virtual & in-person: In some instances, I have used a hybrid approach to neuropsychological assessment of clinic patients. This typically will involve virtual interview (i.e., via secure virtual platform) along with in-person cognitive testing and in most cases online psychological testing.

Regardless of format, consideration of a multitude of factors in the presentation of cognitive concerns is a critical component of neuropsychological evaluation. Balancing the importance of such factors within the context of a neuropsychologist's clinical judgment is illustrated below (Fig. 2.2).

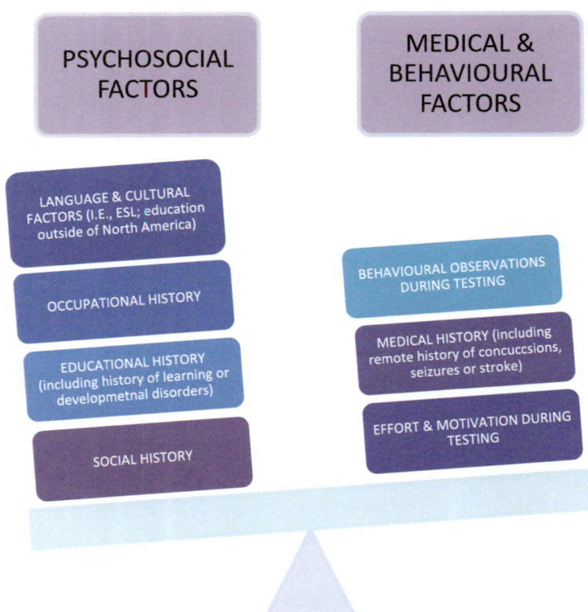

Fig. 2.2 Factors for interpretation of cognitive factors

Analysis of Cognitive Profiles for Diagnosis of Dementia

Following clinical interview along with the comprehensive cognitive and mood testing, the neuropsychologist must analyze and interpret assessment findings within the context of each patient's history and behavioral presentation observed throughout the evaluation. The raw test scores from neuropsychological testing are relatively meaningless without comparison to a normative sample as well as a method for interpreting and communicating the meaning of that comparison (Brooks et al. 2011). As mentioned, the basis of comparison within neuropsychological testing requires understanding of psychometrics (i.e., the study of theory and technique in measurement).

Psychometrics forms the foundation for interpreting neuropsychological assessment by providing the scientific framework of reliability (i.e., consistency of results) and validity (i.e., measuring what it claims to measure). This allows for meaningful clinical interpretations of test scores by ensuring the tests are standardized, free of measurement error, and accurately reflect an individual's performance compared to a normative group, enabling reliable diagnosis and tracking of cognitive functions. The psychometric principles considered within neuropsychological interpretation include the following:

- *Reliability*: A neuropsychological test must be reliable, meaning it consistently produces similar results under the same conditions. This minimizes the influence of random error and ensures that the observed performance is a true reflection of the individual's ability.
- *Validity*: A test's validity determines whether it accurately measures the specific cognitive ability it is intended to assess. This is crucial for making accurate clinical inferences, such as detecting the presence or absence of cognitive dysfunction.
- *Standardization*: Psychometrically sound tests are standardized, which means they are administered and scored in a uniform way. This creates an objective and repeatable process that allows any qualified neuropsychologist to obtain comparable results.
- *Predictability*: A psychometrically acceptable procedure is predictable; so that its results can be expected to occur in predictable patterns based on research studies, allowing for reliable and meaningful clinical inferences.

With this psychometric foundation, the neuropsychologist will analyze patterns of cognitive performance both between cognitive domains (i.e., patterns of memory vs attention etc.) and within cognitive domains (i.e., patterns of performace within multiple verbal memory tasks). The interpretation and communication of neuropsychological test results "depends upon having an appropriate comparative sample and a common "language" of descriptors (i.e., classification system). Regardless of the classification system that is used, it is recommended clinicians be consistent throughout a report" (Brooks et al. 2011). The analysis of patterns of cognitive performance within this context allows the neuropsychologist to come to opinion regarding diagnosis and management (Fig. 2.3).

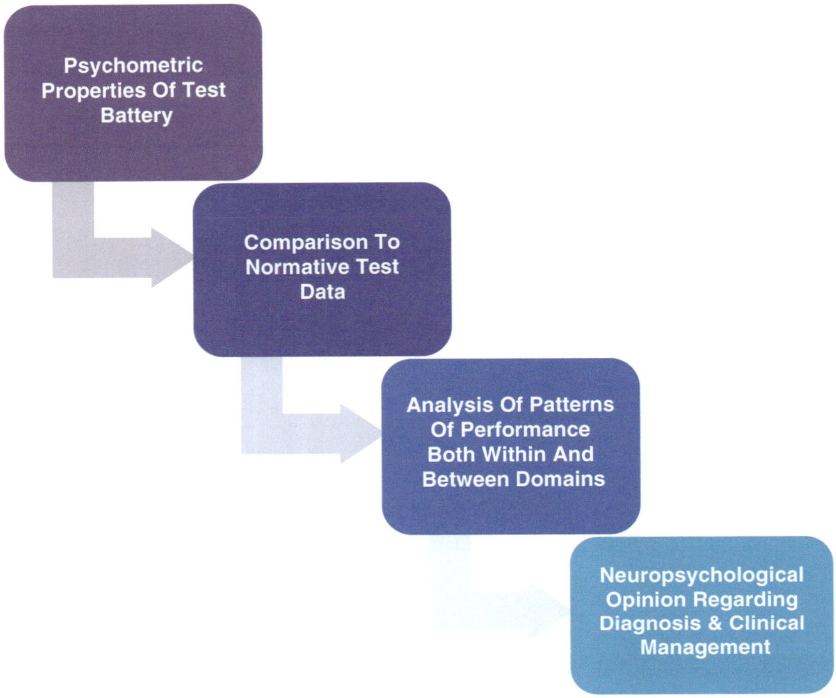

Fig. 2.3 Process of neuropsychological interpretation

Longitudinal Follow-Up of Cognitive Profiles for Diagnosis of Dementia

With respect to diagnostics in dementia, and certainly within my own clinical practice, perhaps one of the most important aspects of assessment of cognition is longitudinal follow-up. Typically, it is the follow-up assessment(s) that may be most informative when diagnostic complexity is present. Specifically, a second or perhaps even third or fourth neuropsychological evaluation can provide great insight into neurodegenerative disease-specific patterns over time.

As discussed previously, there is variance in predictive models for conversion from prodrome to diagnosis of dementia. Depending upon criteria used to define MCI, it is believed that approximately one in four persons diagnosed with MCI progress to dementia after 3–5 years (McGrattan et al. 2022). There is also considerable variability in *how* individuals progress or do not progress to dementia. Of course, it is important to acknowledge that not all who are labeled MCI will progress to dementia, and these progressions occur within the context of normal aging, which involves a degree of cognitive change over time. Notably, changes related to normal aging, which may start in the 30s, typically include subtle decline over time although these changes do not significant impede daily functioning.

However, in general, the cognitive areas most likely impacted by *normal aging* are as follows:

Language: With normal aging, our vocabulary seems to remain relatively stable or may improve over time particularly in those who engage in lifelong learning. However, naming (i.e., our ability to produce the labels) and fluency (i.e., our ability to express ourselves through production of words in a timely manner) declines somewhat with advancing age.

Visual-Spatial Abilities: In general, visual-spatial abilities (i.e., spatial perception as well as visual recognition and object perception) are well preserved with age. However, visual construction (i.e., ability to construct a visual puzzle) declines over time.

Processing Speed: Our speed of cognitive performance begins to decline in the 30s and continues to decline over subsequent decades. This cognitive slowing can impact other areas of function across various cognitive domains, including memory (i.e., given impact of slower processing on the encoding of incoming information).

Executive Functions: Many aspects of complex executive function, such as reasoning and problem solving with familiar information, remain stable while we age. However, there is a decline in response inhibition (i.e., brain's ability to stop or pause an impulse or behavior in process), inductive reasoning (i.e., making an educated guess based upon details or information gathered), and mental flexibility with advancing age.

Memory: Table 2.2 summarizes memory function to distinguish aspects expected to decline with age and those that remain unchanged with advancing age (Harada et al. 2013; Murman 2015). As fear of forgetting is the most frequent concern for older adults, this distinction is critical (Table 2.2).

Overall, although cognitive changes as described above occur with normal aging and can be concerning for individuals as they age, progression to MCI or dementia

Table 2.2 Memory changes with age

Memory decline with age	Unchanged memory in aging
Involuntary memory—the spontaneous retrieval of material without use of cues (i.e., *recall of shopping list*)	*Recognition memory*—the recall of previously learned material with aide of cues or reminders
Source memory—recalling the source (what when or how) information was learning (i.e., *forgetting that you heard information on the news rather than article*)	*Temporal memory*—the ability to remember the timing and sequence of events (i.e., *remembering you had breakfast before the meeting*)
Prospective memory—remembering to perform activities in the future (i.e., *taking daily medication; attending planned appointments*)	*Procedural memory*—involves the implicit (unconscious) recall of how to perform skilled actions or habits (i.e., *how to ride a bike*)
Episodic memory—the ability to recall specific events or experiences (i.e., *remembering what you were doing at the time of 9/11*)	*Remote memory*—the ability the remember information and events learning in the remote past (i.e., *remembering your 16th birthday*)

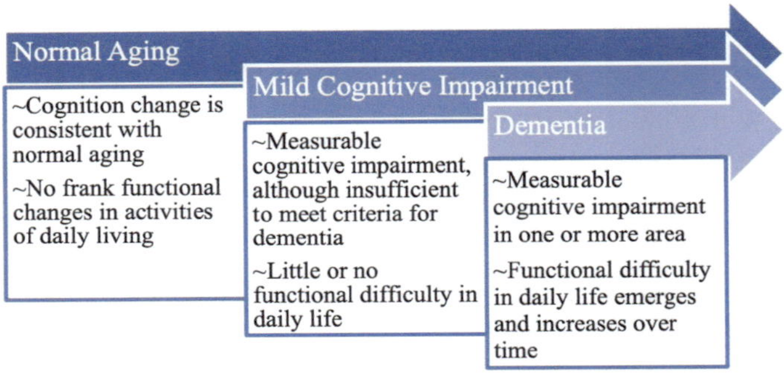

Fig. 2.4 Progression of cognition and function

status reflects decline beyond this "normal aging" level as reflected within cognitive testing. Neuropsychological evaluation, which is normed for age (i.e., so performance is compared to those of equivalent age), provides a clinically useful means of differentiating those aspects of cognitive decline expected with normal aging versus the possibility of an emerging neurodegenerative condition.

Figure 2.4 provides a simplified view of the cognitive and functional progression from cognitive aging through MCI and dementia (Fig. 2.4).

As mentioned, of particular clinical significance, is the cognitive *variations* that may present themselves within and across neuropsychological assessments in longitudinal follow-up. Both within-domain variability (i.e., variance seen in tests within the same cognitive area) and across-domains variability (i.e., pattern of variance across different cognitive areas) are useful when mental health factors coexist within the context of cognitive concern. The frequency of neuropsychological reassessments may vary depending on presentation although the image below reflects my triage process in the Clinic for Alzheimer's Disease and Related Disorders. The triage process depicted represents my own practice and may vary depending on clinician and clinic requirements (Fig. 2.5).

The importance of triage cannot be underestimated for longitudinal follow-up within the diagnosis of dementia, particularly for those with coexisting mental health issues. In my own practice, I have found educating patients and their families on my approach to the triage process is an asset to minimize any frustration or misunderstandings should there need to be extended time between assessments. This process of triage is essential to minimize negative consequences that may be associated with clinical neuropsychological evaluations such as:

- Distress associated with anticipatory anxiety regarding cognitive testing, particularly in those individuals who may have history of test-taking anxiety or learning disabilities.
- Increase in rumination and negative perception regarding past neuropsychological performance in those individuals with past cognitive assessments.

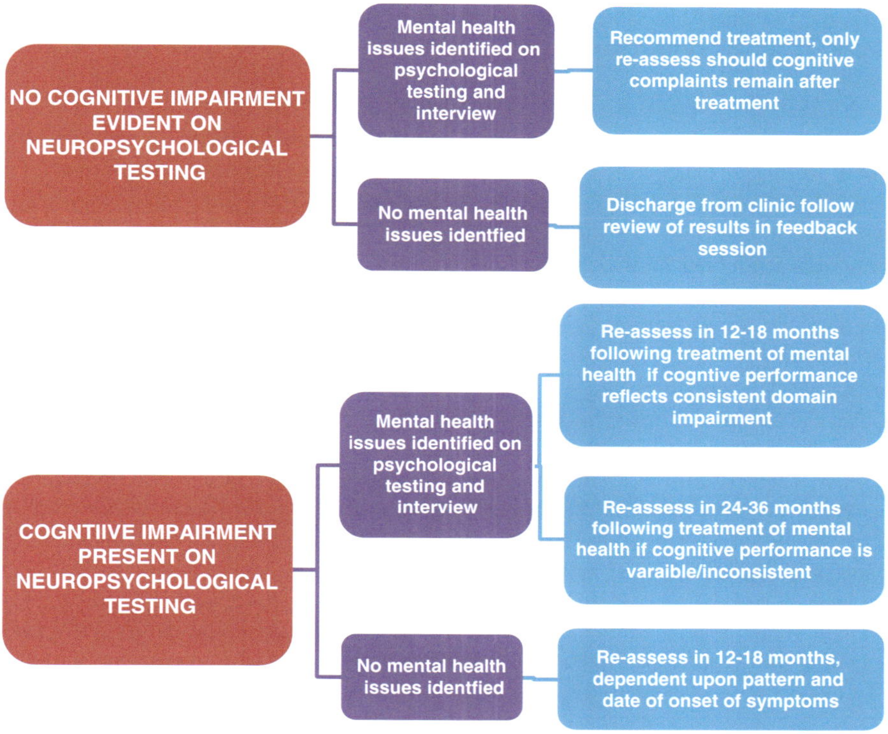

Fig. 2.5 Triage longitudinal neuropsychological assessment

- Performance anxiety consequent to the feeling of lack of preparation, which then can lead to increased levels of stress and shame.
- Mental fatigue during and following assessment, however transient, which may occur with full neuropsychological testing (i.e., often entailing 4–6 hours) can lead to increase in cognitive errors and thus exacerbate mental health factors (i.e., anxiety and misperceptions regarding performance).

As will be discussed in subsequent chapters, the very process of evaluation in dementia can lead to heightened mental health factors, particularly anxiety for a portion of patients presenting to memory clinics. Cognitive testing can, indeed, be emotionally intrusive for some patients, who may have awareness of failures as well as those who may overestimate or underestimate their performance during the testing process. I continue to have many patients who express considerable anticipatory anxiety prior to engaging in testing (i.e., "I am so nervous about this") and or heightened reactivity to their perceived performances during testing (i.e., "I am failing all of these tests"). It is not uncommon for my neuropsychological technician to document a patient's severe reactivity to some, or all of the testing conducted. This can be an important clue to how such patients perceive and/or react to their own cognitive capabilities. Communications regarding such behaviors and perceptions are an important component of clinical care and will be discussed in more detail in subsequent chapters.

Further, many patients I see for neuropsychological assessment are observed to ruminate regarding future outcomes of their cognitive performance which may serve to further exacerbate issues of performance during formal evaluation (i.e., "I know I am going to fail these tests") leading to poor performance related to anticipatory anxiety/distress. In the current time of confirmation biased self-directed research conducted by patients, (i.e., 'I googled this' or "I did my own research") this can lead to increased mental health symptoms over time. I have seen many patients on neuropsychological reassessment who express detailed recall of their prior cognitive test performances from 12 to 18 months earlier, at the time of second evaluation. Even upon initial consultation, clues to such hypervigilance can be observed and provides useful insight into patient perception and thus, may inform diagnostic opinion.

One such patient was a 48-year-old woman we can call Janet, who upon initial assessment interview described her considerable concern for memory loss. Within this context, Janet proceeded to describe, in great detail, every instance of memory lapse she had experienced within the past 2 years. The degree of detail she was able to provide was certainly a clue to her hypervigilance which was reflected on psychological testing (i.e., anxiety and obsessive-compulsive traits endorsed on self-report mood testing) despite relative preservation of cognitive performance. Following feedback for her initial neuropsychological assessment regarding these intact performance outcomes, Janet persisted with believing her perceived cognitive issues were attributable to an emerging dementia and so she did not pursue mental health treatment as I had recommended. Upon reassessment approximately 2 years later, Janet's anxiety, hypervigilance, and rumination regarding perceived cognitive change had escalated further although her cognitive performances on neuropsychological testing remained intact with minor variability. It was within this second neuropsychological feedback on results that Janet was able to accept the possible role of anxiety in her functional cognitive issues and she was more amenable to mental health treatment.

I often consider the functional struggles such distress must cause for patients like Janice and how this distress can serve to give the illusion of cognitive decline for such individuals. Thus, it is important for such patients to have mental health treatment in the interim between neuropsychological evaluations for dementia investigation in order to better manage such ruminations. Without interim mental health treatment, particularly in the context of increasing mental health symptoms, cognitive deficits may also appear to increase which then leads to misinterpretation of cognitive decline by the patient and the clinician. This misinterpretation may lead to misdiagnosis and inappropriate or delayed treatment for such individuals, as suggested by the illustration below (Fig. 2.6):

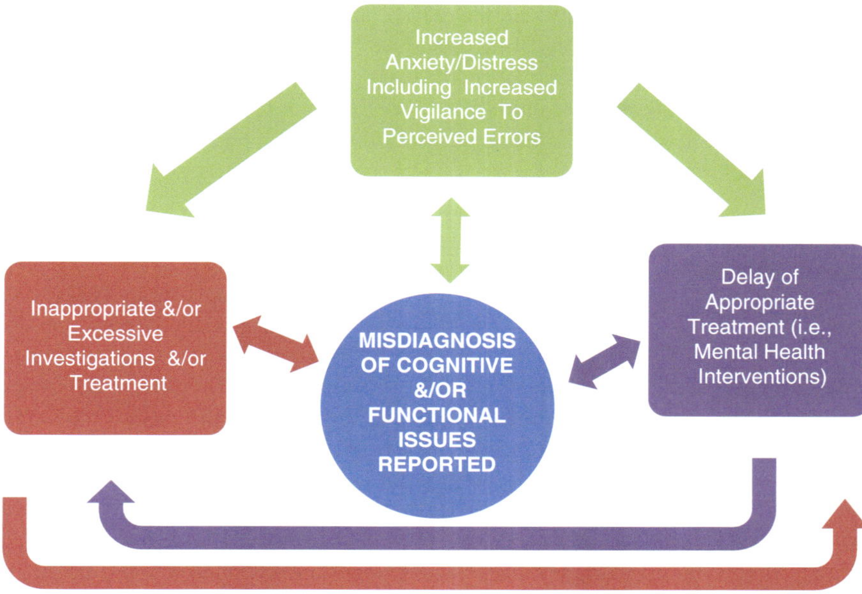

Fig. 2.6 Misdiagnosis of cognitive complaints

As illustrated above, there are multidirectional relationships between a misdiagnosis of cognitive complaints and its various consequences, including the following:

- Anxiety and vigilance may increase in the context of misdiagnosis of symptoms and/or diagnostic uncertainty.
- Inappropriate and/or excessive clinical investigations and/or treatments may be pursued, particularly in patients who are anxious who may persistently request further investigation due to perceived decline.
- Appropriate treatment, such as mental health interventions, is often delayed in the context of misdiagnosis of diagnosis when mental health may be a significant contributor to cognitive change.

This multidirectional impact of misdiagnosis and diagnostic uncertainty increases clinical complexity and healthcare costs in such cases, along with the considerable distress for patients. Such distress may include increased symptoms of anxiety, depression and/or functional cognitive symptoms as will be discussed in future chapters. Underestimating the impact of such distress may serve to hinder diagnostic specificity particularly in prodromal or early stages of neurodegenerative disease, given the heightened risk for mental health disorders consequent to such distress.

It is important to recognize the report and/or perception of cognitive impairment may be distorted within a subset of patients presenting for clinical dementia assessment. This disconnect of perception regarding cognitive capacity and measured cognitive performance is one of the cornerstones of recognizing functional cognitive disorders to be discussed in detail in later chapters. Research suggests an over-reliance on subjective report of cognitive concerns within the prodromes of dementia may lead to increased likelihood for misdiagnosis (Edmonds et al. 2014). These researchers found no significant relationship between reported cognitive symptoms and cognitive performance on testing. Further, the researchers conclude the following:

> "The inclusion of self-reported complaints in MCI diagnostic criteria may cloud rather than clarify diagnosis and result in high rates of misclassification of MCI. Discrepancies between self- and informant-report demonstrate that overestimation of cognitive problems is characteristic of normal aging while underestimation may reflect greater risk for cognitive decline."

Within clinical practice, I have found this pattern of overestimation versus underestimation of self-reported cognitive symtpoms to be one of the initial indices for analysis of the cognitive and psychological testing conducted with the patients I see, as was demonstrated in the case of Janet discussed earlier. In other words, overestimation of cognitive impairment by patients may be an initial indicator of mental health factors within the clinical presentation. Conversely, in patients who underestimate their cognitive impairment it is not uncommon to find marked cognitive impairment reflected on neuropsychological testing. Obviously, diagnostic opinion should never rely on such patterns of perception alone although they provide useful clinical insights. In fact, as will be discussed in future chapters, patients' misperceptions regarding cognition may provide valuable clues to underlying mental health conditions, such as functional cognitive disorders (FCD). Nevertheless, validating a patient's *experience* of symptoms is important regardless of cause as being dismissive may increase the likelihood for avoidance of appropriate mental health treatment even when strongly indicated. Patients with heightened anxiety and vigilance may be more likely to have their cognitive concerns dismissed within clinical settings, even for those whose cognitive impairment prove to be progressive over time consequent to an emerging dementia. Without comprehensive clinical evaluation (i.e., neuropsychological evaluation), diagnostic specificity regarding etiology of cognitive symptoms and the perception of cognitive decline can be limited.

Overall, the nuance of cognitive decline or variability in patients presenting with subjective cognitive complaints in the context of evaluation for possible dementia requires the in-depth analysis of both inter- and intra-domain cognitive performance. Although patterns of cognitive performance within various forms of dementia will not be reviewed here, the patterns present within the complex interaction between mental health factors and cognition will be discussed in depth within later chapters of this book. Within the context of memory clinics focussed upon dementia diagnostics, the understanding and comprehensive evaluation of cognitive patterns for all variants of neurodegenerative and mental health disorders is critical in order to mitigate possible misdiagnosis and mismanagement of symptoms. Furthermore, comprehensive neuropsychological evaluation of cognitive and mental health symptoms may serve as a conduit for effectively communicating challenges to patient's misperceptions regarding their cognitive skills as is evident in disorders such as functional cognitive disorder (FCD), to be discussed in future chapters.

References

Basagni B, Navarrete E. Suggestions for the decision making in subjective cognitive complaints. Aging Clin Exp Res. 2025;37(1):27.

Brooks B, Sherman E, Iverson G, Slick D, Strauss E. Psychometric foundations for the interpretation of neuropsychological test results. In: The little black book of neuropsychology: a syndrome-based approach. New York: Springer; 2011. https://doi.org/10.1007/978-0-387-76978-3_31.

Brown T, Zakzanis KK. A review of the reliability of remote neuropsychological assessment. Appl Neuropsychol Adult. 2025;32(5):1536–42.

Cabreira V, Alty J, Antic S, Araújo R, Aybek S, Ball HA, Baslet G, Bhome R, Coebergh J, Dubois B, Edwards M, Filipović SR, Frederiksen KS, Harbo T, Hayhow B, Howard R, Huntley J, Isaacs J, LaFrance WC Jr, Larner AJ, Di Lorenzo F, Main J, Mallam E, Marra C, Massano J, McGrath ER, McWhirter L, Moreira IP, Nobili F, Pennington C, Tábuas-Pereira M, Perez DL, Popkirov S, Rayment D, Rossor M, Russo M, Santana I, Schott J, Scott EP, Taipa R, Tinazzi M, Tomic S, Toniolo S, Tørring CW, Wilkinson T, Frostholm L, Stone J, Carson A. Perspectives on the diagnosis and management of functional cognitive disorder: an international Delphi study. Eur J Neurol. 2025;32(1):e16318.

Edmonds EC, Delano-Wood L, Galasko DR, Salmon DP, Bondi MW, Alzheimer's Disease Neuroimaging Initiative. Subjective cognitive complaints contribute to misdiagnosis of mild cognitive impairment. J Int Neuropsychol Soc. 2014;20(8):836–47.

Harada CN, Natelson Love MC, Triebel KL. Normal cognitive aging. Clin Geriatr Med. 2013;29(4):737–52.

McGrattan AM, Pakpahan E, Siervo M, et al. The DePEC Team: Risk of conversion from mild cognitive impairment (MCI) to dementia in low-and middle-income countries (LMICs): a systematic review and meta-analysis. Alzheimers Dement. 2022;8:e12267.

Murman DL. The impact of age on cognition. Semin Hear. 2015;36(3):111–21. https://doi.org/10.1055/s-0035-1555115.

3 The Importance of Sex-Specific Risks

In the early years of my career, despite being a woman, sex-specific differences in cognition and risk for neurological or psychiatric issues for my patients were not prominent in my mind. However, with the experience of hormone-based challenges to my own brain and cognitive function, I began to acknowledge the distinct changes in the way a women's brain functions throughout her lifecycle. During my first pregnancy at age 39, I experienced the emergence of word-finding difficulties, distractibility, and forgetfulness which were perhaps more functionally prominent from my perspective given the nature of my work. Thus, began my burgeoning interest in the endocrine system and sex-based differences in cognition.

It was during this period of my clinic practice that I noticed a shift from patients solely meeting diagnostic criteria for dementia toward a younger demographic of patients presenting primarily with "worry" about cognitive change despite falling short of any dementia criteria. Specifically, I was receiving an increasing number of referrals for women in their late 40s and early 50s with complaints of perceived-emerging cognitive issues and decline, particularly in the areas of memory and word-finding skills. My own experience with similar cognitive challenges in pregnancy heightened my awareness of the possible impact of hormonal changes during transition to menopause and beyond. This then became a more prominent consideration in my neuropsychological formulations of cognition in such cases despite minimal clinical research in the area at that time. Thankfully since then, the body of research regarding the impact of hormones on women's cognitive function is growing, albeit at a frustratingly slow pace. Let us take a closer look at evolution of some of this research.

Sex-Based Disparity in Research

A discussion of sex-specific risks in the diagnosis of dementia, including any possible co-exiting mental health issues, must first acknowledge the history of sex-based disparity in research and consequently reflected in clinical care. Notably, the

S. Hayden, *Fear of Forgetfulness*, https://doi.org/10.1007/978-3-032-21818-6_3

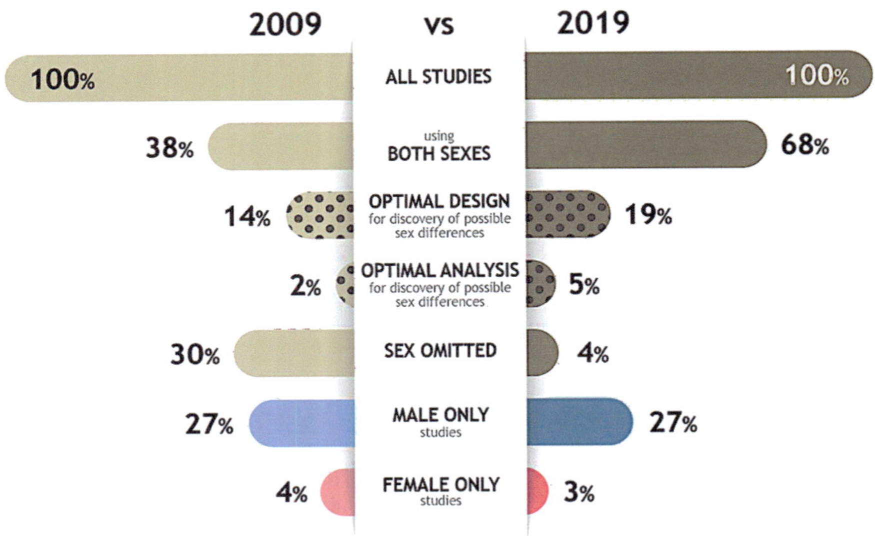

Fig. 3.1 Sex-based disparity remains in research (Rechlin et al. 2022)

significant underrepresentation of females in research studies, even within animal studies, persisted until the 1990s when NIH legally and ethically mandated the inclusion of women in research. Specific to neurological and psychiatric population studies, this sex-based discrepancy was well illustrated by a group of University of British Columbia researchers (Rechlin et al. 2022) who examined how biological sex is incorporated into the design and analysis of published literature within neuroscience and psychiatry journals between 2009 and 2019. Although this review encouragingly found a 30% increase in the number of papers reporting studies including both sexes in 2019 compared with 2009, there was no change in the number of studies that included females only (3%) relative to male only studies (27%) in this 10-year span. The researchers rightfully concluded there is need for further study of sex differences within neurological and psychiatric populations (Fig. 3.1).

Neurological and Psychiatric Sex-Based Risk

Unfortunately, this sex-based disparity in research has had, and continues to have, considerable impact on evidence-based clinical care for women and in how informed women are regarding their own care (O'Neal, 2024). This research disparity is of particular importance for women with concerns for cognitive decline, given their greater risk for many neurological and psychiatric conditions. Research suggests women with mild cognitive impairment (MCI) experience declines in cognition and function at nearly two times the rate of men (Lin et al. 2015). Other scientists note that a 45-year-old woman has a one in five chance of developing Alzheimer's disease during her lifetime versus a one in ten chance for same age males (Moutinho 2025), although the reason for this sex-based difference remains unclear. This is heightened sex-specific risk factor is complicated further by the fact that caregivers of elders are largely women, estimated to be 57%–81% depending on the country (Sharma et al. 2016). The stress associated with caregiving therefore disproportionally impacts women and further potentially increases risk for cognitive decline and/or mental health conditions (Christian et al. 2023).

Challenges also exist in the pattern of cognitive presentation for women versus men, with females showing a cognitive advantage in early Alzheimer's disease, albeit with women experiencing more rapid decline in later stage illness (Sundermann et al. 2019). Notably women exhibit stronger verbal memory, which may delay or distort MCI identification, resulting in a "masking effect" with evidence of 20% diagnostic errors when using non-sex-adjusted verbal memory norms. Specifically, this 20% of diagnostic errors was reflected in 10% false negatives for women (i.e., underdiagnosis) and 10% false positives for men (i.e., overdiagnosis). The science of this trajectory was reported by Emrani and Sundermann (2025) who found women to be better able than men to sustain normal cognitive performance, particularly verbal memory, in the presence of early-stage Alzheimer's disease pathogenesis. However, once the disease pathology burden reaches a tipping point beginning in the moderate stages of illness, women can no longer compensate, resulting in more rapid cognitive decline compared to males.

Further of note, research suggests there are differences for women versus men in biomarker risk with APO-E 4 (i.e., the susceptibility gene variant that raises risk for Alzheimer's disease) conferring greater risk of Alzheimer's disease for women than in men (Holland et al. 2013).

Heightened brain health risks within other neurological conditions for women relative to men also exist. For example, Carr and associates found that women experience concussions more frequently suffer more symptoms (particularly if sustained in luteal phase of menstrual cycle) and can take longer to recover than men (Carr and Fleddermann 2025). These findings suggest the possibility that luteal phase hormonal fluctuations may play a role in the worse clinical concussion outcomes evident for women and reflects the need for considering such factors in the diagnosis and management of concussions. These findings do not suggest concussion will result in cognitive decline for women over time, although understanding women's risk for concussion is important in patients presenting with concern for cognitive decline in context of concussion history, even if such concussion events occurred remotely. Significant symptoms of concussion may persist for years, depending on severity of concussion. When such cognitive symptoms are superimposed upon normal aging later in life, this can give an illusion of cognitive decline mimicking a neurodegenerative process.

Unfortunately, more negative clinical outcomes for women also exist for cerebrovascular events (i.e., stroke), with women having higher lifetime risk and worse stroke outcomes than men (Yoon and Bushnell 2023). Other researchers suggest the worst outcomes in women may be attributable to age and other comorbidities in women although the basis for sex differences requires more research (Carcel et al. 2020).

Additional risk includes women estimated to be 22%–45% more likely to have symptoms of long COVID (i.e., symptoms that persist for over 4 weeks following initial infection), with highest risk for women aged 40–55 (Shah et al. 2025). These researchers also found that prior literature combined with their own data "suggests that differences in hormonal levels may partially explain the higher prevalence of long COVID in females younger than 55 years. An attenuation in the risk ratio of long COVID for females aged 18–39 years may be explained by most pregnant individuals belonging to this age category." Sex differences also exist in the clinical presentation of long COVID as well, with women at a higher risk for long-COVID neurological symptoms (Gorenshtein et al. 2025). Once again, further research on sex differences in this area is required.

Sex differences are also evident within many psychiatric conditions, with women being twice as likely than men to be diagnosed with various mental health conditions, including anxiety, depression, and post-traumatic stress disorder (PTSD) (Tannenbaum and Boksa 2019). Further, Sialino and associates (2020) found "a consistent female disadvantage in mental health across different sociodemographic groups and over decennia (1992 vs. 2002) with no specific risk groups."

As noted with concussion, understanding other aspects of a woman's neurological and mental health history which may have associated cognitive concerns is critical when considering cognitive decline in later life given the potential for such issues to mimic dementia when superimposed upon normal aging. This is particularly relevant for women who may also experience cognitive shifts related to perimenopause and beyond. Overall, significant sex differences for risk for both neurological and psychiatric conditions have been found in literature to date, although more research is certainly required to further explore the etiology and outcomes of such differences. Such exploration will serve to better inform diagnostics and management/treatment for neurological and mental health issues in women for the future.

Sex-Specific Brain Function and Cognition

One of the pivotal research findings in sex-based neuroscience occurred in the 1980s, when estrogen receptors concentrated in the hypothalamus, pituitary hippocampus, mid brain, cerebral cortex, and brainstem were identified and mapped (Sherwin 2003). This finding acted as a catalyst for the neuroanatomical understanding of functional correlates in affected brain regions and how they may be impacted by hormonal fluctuation and change in women. Although an in-depth review of neuroendocrinology is beyond the scope of this book, below is a rudimentary review of the current understanding of sex differences in brain function and cognition to provide perspective of sex-based commonalities and diversity.

While the brains of women and men are more alike than they are different, there is growing evidence for sex-based differences in the hormonal, structural, and, to some degree, functional operations of our brains. For instance, males in general have brains that are approximately 10% larger than women's brains (Sowell et al. 2006). Despite varying findings across this research landscape, there is no clear or consistent evidence that brain size impacts intelligence between the sexes; however, it may be implicated in the differences in the approach to cognitive processing between women and men. An often used analogy suggests the pathway or route to a final destination for a cognitive skill may differ for men and women although they will both arrive at the same destination.

Studies determined men generally have a larger inferior parietal lobule versus women and it is proposed that the larger size of this brain structure is related to stronger mathematic skills and increased accuracy with time and speed estimations. Some neuroimaging studies demonstrate the hippocampus, associated with learning and memory, is larger in women than men when adjustments are made for overall brain size (Cahill 2006). However, more recent studies do not consistently support this finding. There is also conflicting evidence suggesting the amygdala (particularly the left amygdala) is significantly larger in men than in women when adjusted for brain size. The variance in outcomes of these studies on neuroanatomical sex differences further support the need for ongoing research not only to explore sex-based differences but to determine what those differences mean clinically in order to facilitate optimal care for both men and women in the future.

Neuroscience does suggest women's brain connectivity, which promotes multi-tasking skills, is more robust than that of men, as women have ten times more white matter than men. Women also have greater blood flow through specific regions that support behaviors related to empathy, collaboration, and intuition. In contrast, men have been found to have 6.5 times more gray matter than women, and this is thought to account for task-focused behavior. Shiino et al. (2017) found subregional sex differences, such as white matter in the bilateral anterior frontal regions and the left lateral white matter to be larger in women. Others have uncovered evidence that connectivity is greater between the front and back of the brain for men, which aids eye-hand coordination and understanding spatial relationships. In contrast, women have greater left to right connectivity promoting stronger skills in pulling information from multiple sources (Ingalhalikar et al. 2014).

A 2006 study by Cahill provided insight into persistent misconceptions about sex-based differences in neuroscience, along with a review of the neurological basis of these differences. He opined that: "sex differences exist in every brain lobe, including many cognitive regions such as the hippocampus, amygdala and neocortex" with many of these differences more evident in the "functional dimension" rather than overt anatomical structure. However, there have been limits in defining sex-based cognitive differences in a consistent manner in subsequent research. Irrespective of political or social implications, it is important to identify sex-based anatomical, hormonal, and functional (i.e., cognitive and emotional) brain differences.

In general, from a cognitive perspective, female-typical skills include women excelling at verbal skills, memory, fine motor skills, and perceptual speed and accuracy. Male-typical skills are reflected in visual memory, mathematical skills, and spatial abilities. These differences have been found to be modest albeit consistent

(Sherwin 2003) in available research. The evolution of sex-based differences in cognition is theoretically believed to be organized by sex differences via exposure to sex hormones during fetal development and later expressed after puberty's introduction of high levels of the same hormone. The hypotheses being that sex hormones impact the brain and possibly sex-specific cognitive abilities (Kheloui et al. 2021). The basis of this neuroendocrine theory is beyond our scope here, but we will learn more about sex-specific cognitive issues in subsequent chapters. The impacts of gender roles and sexual orientation in brain and cognitive function also need to be examined to generate more rigorous research and to support precision care.

Recent research (Zuhlsdorff et al. 2026) has also explored the impact of menopause on brain matter volume, with menopause being linked to increase in various mental health factors with evidence for structural brain changes in some areas. Specifically, this study revealed postmenopausal women had smaller medial temporal lobe (MTL—involved in memory and emotional regulation; highly impacted in Alzheimer's disease) and anterior cingulate cortex (ACC—involved in executive function) volumes relative to premenopausal women. Such findings provide preliminary clues to the possible mechanisms of sex-based differences for neurodegenerative disease and mental health issues.

Overall, the existence of such sex-based brain differences continues to be reflected in the scientific literature and thus, the clinical applications of this knowledge in healthcare remains limited. It is important to assert there are no sex-based differences in overall cognitive capacities despite varying research to support sex-based differences within some cognitive domains. As mentioned previously, men and women likely take different paths to arrive at the same destination, when performing specific cognitive functions.

Ignoring sex-based differences in patients with neurological or mental health issues does not advance or support optimal care for either male or female patients. A patient's sex and gender identity are critical to their lived experience of symptoms in clinical care. Should clinicians choose to ignore sex-based factors in cognitive and mental health functioning, there is risk of missing critical clues related to precision diagnostics and for optimized management of clinical symptoms.

I have seen these issues many times in my own clinical practice of neuropsychology. It is common during my clinical interview of female patients for them to reveal significant psychosocial and/or possible hormonal factors influencing their presentation of cognitive or mood symptoms (e.g., reporting memory change which coincides with self-described transition to menopause or during time of high perceived stress). For example, a woman acting as primary caregiver of her parents with dementia is more likely to hyper-focus upon her own cognitive issues during times of stress. This may possibly be further compounded by hormonal factors related to the transition to menopause and beyond. Without asking about and understanding such factors that are more prevalent for women, symptoms of forgetfulness may be misinterpreted from a clinical perspective.

I see many women in or around their 50s who come to me in considerable distress with complaints of emerging and possibly worsening memory issues which are later found to be related to attentional factors consequent to stress and/or menopausal symptoms such as lack of sleep or hormonal change. When considering the onset and or persistence of cognitive symptoms, I now routinely include questions related to women's hormonal status (e.g., menstrual change following brain injury; menopause status when concerned with neurodegenerative conditions) in my effort to provide a more thorough clinical opinion. This approach to care is critical despite the paucity of research providing clear understanding for the mechanisms underlying sex-based differences in cognition and potential cognitive decline. Providing women with education regarding the state of research in this area, including limitations of such research, can support their understanding of the factors that may impact cognitive performance for them.

I also have seen many women in my clinical practice who describe their experience of a medical system that repeatedly attributes cognitive symptoms to mood and stress issues as opposed to a neurological disorder, or dismisses their concerns all together with minimal explanation, only to later be diagnosed with an emerging dementia. Such dismissal or minimization of women's cognitive symptoms may relate to the sex-based differences in presentation and course of neurodegenerative disease discussed earlier, although this may also reflect other sex-based biases within the research and medicine (Aggarwal and Mielke 2023).

Thus, discussing concern regarding cognitive function can feel daunting for many women as it can be difficult to express these concerns in a concise manner during the brief time most family physicians have for appointments. There is some evidence for sex-based differences in the reporting of cognitive symptoms to family physicians. A Cleveland Clinic survey for the Women's Alzheimer's Movement (2022) revealed 73% of women do not discuss their brain health with their doctor. This survey also suggested there is a disconnect between women's experience of cognitive and mental health symptoms and their understanding of how they may relate to hormones or other sex-based factors. Although 71% of women surveyed had seen a physician in the past year, it seems that appropriate conversations about their cognitive and mental health symptoms did not happen. As a neuropsychologist, I certainly understand my role in advocating for such discussions within my evaluations. However, neuropsychologists are rarely the first point of contact when such symptoms become worrisome for women. Therefore, women should be encouraged to first communicate their perceived symptoms with their primary physicians. Women deserve more education regarding how cognitive and mental health symptoms reflect their overall brain health, and this increased knowledge is a positive step toward being able to advocate for precision healthcare management in their own lives.

In summary, understanding sex-based differences in risk and cognitive presentation for neurological and psychiatric conditions is critical for thorough diagnostic opinion and treatment recommendations, particularly for women. To optimize clinical care for women in this area, development of sex-specific norms for cognitive testing would ensure more accurate disease detection and would facilitate more effective sex-specific treatment options. As with all patients, regardless of sex or gender, optimal clinical care requires enhanced understanding of all possible mitigating factors in the presentation of symptoms. Emerging research suggests sex-based differences in risk and presentation may be an important mitigating factor for both neurodegenerative and mental health disorders. Consequently, to advance scientific understanding and precision clinical care within memory clinics specialized in dementia diagnostics, inclusion of sex-based differences will be essential.

References

Aggarwal NT, Mielke MM. Sex differences in Alzheimer's disease. Neurol Clin. 2023;41(2):343–58.

Cahill L. Why sex matters for neuroscience. Nat Rev Neurosci. 2006;7:477–84.

Carcel C, Woodward M, Wang X, Bushnell C, Charlotte Sandset E. Sex matters in stroke: a review of recent evidence on the differences between women and men. Front Neuroendocrinol. 2020;59:100870.

Carr G, Fleddermann MT. The influence of menstrual cycle phases on postconcussion outcomes and symptom reporting: a scoping review. Scand J Med Sci Sports. 2025;35(7):e70093.

Christian LM, Wilson SJ, Madison AA, Prakash RS, Burd CE, Rosko AE, Kiecolt-Glaser JK. Understanding the health effects of caregiving stress: new directions in molecular aging. Ageing Res Rev. 2023;92:102096.

Cleveland Clinic News Release. The women's Alzheimer's movement at Cleveland Clinic survey: women are unaware of critical aspects of their health. 23 June 2022.

Emrani S, Sundermann E. Sex/gender differences in the clinical trajectory of Alzheimer's disease: insights into diagnosis and cognitive reserve. Front Neuroendocrinol. 2025;77:101184.

Gorenshtein A, Leibovitch L, Liba T, Stern S, Stern Y. Gender disparities in neurological symptoms of long COVID: a systematic review and meta-analysis. Neuroepidemiology. 2025;59(4):426–40.

Holland D, Desikan RS, Dale AM, McEvoy LK, Alzheimer's Disease Neuroimaging Initiative. Higher rates of decline for women and apolipoprotein E epsilon4 carriers. AJNR Am J Neuroradiol. 2013;34(12):2287–93. https://doi.org/10.3174/ajnr.A3601. Epub 2013 Jul 4. Erratum in: AJNR Am J Neuroradiol. 2015;36(10):E67.

Ingalhalikar M, Smith A, Parker D, Satterthwaite TD, Elliott MA, Ruparel K, Hakonarson H, Gur RE, Gur RC, Verma R. Sex differences in the structural connectome of the human brain. Proc Natl Acad Sci USA. 2014;111(2):823–8.

Kheloui S, Brouillard A, Rossi M, Marin M, Mendrek A, Paquette D, Juster R. Exploring the sex and gender correlates of cognitive sex differences. Acta Psychol. 2021;221:103452.

Lin KA, Choudhury KR, Rathakrishnan BG, Marks DM, Petrella JR, Doraiswamy PM, Alzheimer's Disease Neuroimaging Initiative. Marked gender differences in progression of mild cognitive impairment over 8 years. Alzheimers Dement (N Y). 2015;1(2):103–10.

Moutinho S. Women twice as likely to develop Alzheimer's disease as men—but scientists do not know why. Nat Med. 2025;31:704–7. https://doi.org/10.1038/s41591-025-03564-3.

O'Neal MA. Women and the risk of Alzheimer's disease. Front Glob Womens Health. 2024;4:1324522. https://doi.org/10.3389/fgwh.2023.1324522. PMID: 38250748; PMCID: PMC10796575.

Rechlin RK, Splinter TFL, Hodges TE, et al. An analysis of neuroscience and psychiatry papers published from 2009 and 2019 outlines opportunities for increasing discovery of sex differences. Nat Commun. 2022;13:2137.

Shah DP, Thaweethai T, Karlson EW, Bonilla H, Horne BD, Mullington JM, Wisnivesky JP, Hornig M, Shinnick DJ, Klein JD, Erdmann NB, Brosnahan SB, Lee-Iannotti JK, Metz TD, Maughan C, Ofotokun I, Reeder HT, Stiles LE, Shaukat A, Hess R, Ashktorab H, Bartram L, Bassett IV, Becker JH, Brim H, Charney AW, Chopra T, Clifton RG, Deeks SG, Erlandson KM, Fierer DS, Flaherman VJ, Fonseca V, Gander JC, Hodder SL, Jacoby VL, Kotini-Shah P, Krishnan JA, Kumar A, Levy BD, Lieberman D, Lin JJ, Martin JN, McComsey GA, Moukabary T, Okumura MJ, Peluso MJ, Rosen CJ, Saade G, Shah PK, Sherif ZA, Taylor BS, Tuttle KR, Urdaneta AE,

Wallick JA, Wiley Z, Zhang D, Horwitz LI, Foulkes AS, Singer NG, RECOVER Consortium. Sex differences in long COVID. JAMA Netw Open. 2025;8(1):e2455430.
Sharma N, Chakrabarti S, Grover S. Gender differences in caregiving among family – caregivers of people with mental illnesses. World J Psychiatry. 2016;6(1):7–17.
Sherwin B. Estrogen and cognitive function in women. Endocr Rev. 2003;24(2):133–51.
Shiino A, Chen YW, Tanigaki K. Sex-related difference in human white matter volumes studied: inspection of the corpus callosum and other white matter by VBM. Sci Rep. 2017;7:39818.
Sialino LD, van Oostrom SH, Wijnhoven HAH, Picavet S, Verschuren WMM, Visser M, Schaap LA. Sex differences in mental health among older adults: investigating time trends and possible risk groups with regard to age, educational level and ethnicity. Aging Ment Health. 2020;25(12):2355–64.
Sowell ER, Peterson BS, Kan E, Woods RP, Yoshii J, Bansal R, Xu D, Zhu H, Thompson PM, Toga AW. Sex differences in cortical thickness mapped in 176 healthy individuals between 7 and 87 years of age. Cereb Cortex. 2006;17:1550–60.
Sundermann EE, Maki P, Biegon A, Lipton RB, Mielke MM, Machulda M, Bondi MW. Alzheimer's Disease Neuroimaging Initiative. Sex-specific norms for verbal memory tests may improve diagnostic accuracy of amnestic MCI. Neurology. 2019;93(20):e1881–9.
Tannenbaum C, Boksa P. Sex: a key consideration in understanding the etiology of psychiatric disorders and improving treatment. J Psychiatry Neurosci. 2019;44(6):364–6.
Yoon CW, Bushnell CD. Stroke in women: a review focused on epidemiology, risk factors, and outcomes. J Stroke. 2023;25(1):2–15.
Zuhlsdorff K, Langley C, Bethlehem R, Warrier V, Romero Garcia R, Sahakian BJ. Emotional and cognitive effects of menopause and hormone replacement therapy. Psychol Med. 2026;56:e24.

Part II

Mental Health in Cognition & Dementia

Pseudodementia: Depressive Cognitive Disorders (DCD)

4

The consideration of mental health factors, particularly depression, within the diagnoses of neurodegenerative disorders was suggested by Leslie Kiloh in 1961 who noted this "pseudodementia" mimics the presentation for various neurodegenerative disorders. Although other mental health disorders, to be discussed in later chapters, may also technically fall into the categorization of what had been formerly called pseudodementia, depression has been the most frequently associated condition with this label.

As is reflected throughout scientific literature, the complexity of pseudodementia, now referred to as depressive cognitive disorder (DCD), is further illustrated by the bidirectional relationship between depression and cognition, as depression can be accompanied by cognitive dysfunction as can dementia be accompanied by depression (Brodaty and Connors 2020; Mouta et al. 2023). This bidirectionality may lead to diagnostic complexities within clinical practice, and perhaps more importantly, confusing perception of functional issues for patients regardless of diagnostic category. Educating patients on the cognitive impact of depression is important to limiting negative perceptions or vigilance to cognitive errors, which may in turn increase depressive symptoms further, in the endless cycle reflecting within the image below (Fig. 4.1).

Such issues are further confounded by increased incidence of somatization in aging populations who then may be more likely to present physiological symptoms of depression that can be perceived as signs of physical or cognitive disease rather than mental health issues (Devita et al. 2022). *Somatization* in simple terms is the expression of mental health issues though physical or cognitive symptoms, as in depressive cognitive disorder. Somatization may occur within the normal range of function for most individuals, such as when stress/anxiety results in "stress headaches." However, somatization becomes a clinically significant concern when symptoms become distressing and/or functionally debilitating. This somatization factor is likely a major contributor to the functional cognitive disorders seen in dementia clinics, which will be discussed in depth in Chap. 6 of this book. Regarding this somatization factor in functional cognitive disorders, McWhirter et al. (2020) importantly state the following:

S. Hayden, *Fear of Forgetfulness*, https://doi.org/10.1007/978-3-032-21818-6_4

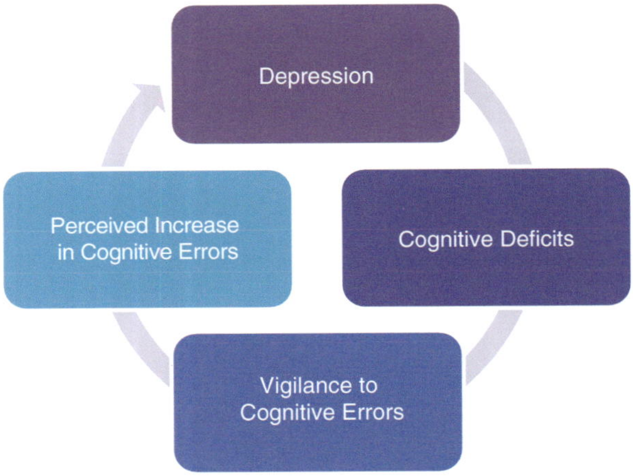

Fig. 4.1 Cycle of depression and cognitive function

> "Around a quarter of patients presenting to memory clinics received diagnoses that might indicate the presence of functional cognitive disorders, which were associated with affective symptoms, negative self-evaluation, negative illness perceptions, non-progressive symptom trajectories, and linguistic and behavioural differences during clinical interactions. Those with functional cognitive disorder phenotypes are at risk of iatrogenic harm because of misdiagnosis or inaccurate prediction of future decline."

As mentioned previously, such harm may be magnified by the assessment process often necessary in the investigative stages of dementia diagnosis. This includes the need for longitudinal follow-up of cognitive performance, which can then lead to persistence and/or exacerbation of functional cognitive issues related to mental health factors. In my own clinic practice, I commonly have patients ruminate on results and/or communications of clinic team members regarding dementia risk which can be exacerbated patients pursuing their own research on the topic. I certainly encourage an empowered and collaborative approach to healthcare with patients and their families, although it is important for clinicians to take the time to understand patient perceptions regarding their clinical presentations in order to mitigate misinformation or misunderstanding of diagnostic information. All these factors require thoughtful and coordinated planning for evaluation and management of cognitive factors in patients with risk for depressive cognitive disorders (DCDs) or for functional cognitive disorders (FCDs) to be discussed in future chapters.

When examining the impact of depression on cognitive function, there is evidence for a positive correlation between the number of depressive episodes and cognitive deficits as well as depression severity and cognitive impairment (Kriesche et al. 2023). This meta-analysis of 42 studies between 2009 and 2019 suggests clinically depressed patients largely show cognitive deficits, relative to health controls, on tests of verbal and visual learning/memory, verbal fluency, processing speed, and some aspects of executive function (i.e., strategy formation and shifting cognitive sets). Given that these cognitive domains are often the same areas found to be impacted within the prodromes of various neurodegenerative conditions, this only emphasizes the need for thorough evaluation of both cognitive and mental health factors in individuals presenting with cognitive complaints within clinical settings.

The etiology of cognitive symptoms in depression is complex and multifactorial, which is well illustrated within literature determining factors that influence cognitive symptoms in major depressive disorder (MDD) as suggested by Zuckerman et al. (2018). The considerable diversity of these factors, including both modifiable and non-modifiable factors, further contributes to the heterogeneity of cognitive symptoms associated with depression. Factors associated with increased likelihood of cognitive symptoms in depressed individuals identified in research are illustrated below (Zuckerman et al. 2018) (Fig. 4.2). As illustrated below the factors to be considered that may impact cognitive symptoms may be longstanding and include issues arising in childhood which requires comprehensive review within clinical interview.

The clinical picture of depressive cognitive disorders is further complicated by research associating late-life depression with increased risk for dementia (Livingston et al. 2020). Other studies have found possible links of amyloid beta to depression as a preclinical symptom of Alzheimer's disease (Smith et al. 2021). Also, higher levels of amyloid beta in cognitive healthy elder adults have been associated with an increased likelihood of developing depression (Donovan et al. 2018). Finally, prevalence of depression is higher in those with mild cognitive impairment (MCI) than in the general older adult population, with the incidence varying widely (i.e., 16%–55%) depending on study design and population (community vs. clinical), but indicates that depression is a common comorbidity that can worsen cognitive decline and increase the risk of progressing to Alzheimer's disease (Ma 2020).

Other research has found depression significantly impairs cognition independent of emergence of a neurocognitive disorder, with estimates of two thirds of patients diagnosed with major depression reporting cognitive dysfunction (Rhee et al. 2024).

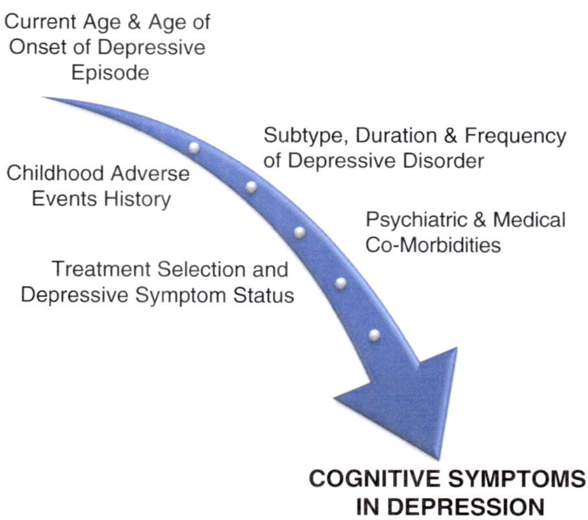

Fig. 4.2 Factors in cognitive symptoms of depression

Such cognitive impairment associated with depression may include measurable change in memory, as reported by Paula et al. (2013) who found memory issues in free recall and learning on a verbally mediated task, although short-term memory and recognition memory is unaffected. Further research suggests depressed patients were found to have deficits in most cognitive domains, notably with cognitive impairment in depression persisting through remission (Rhee et al. 2024). Research has identified specific areas of persistent impairment during remission for depressed patients, including attention/working memory as well as long-term memory (Semkovska et al. 2019). Taken together, all of this research certainly demonstrates the complexity of differentiating depressive cognitive disorders from dementia and its prodromes.

To further conceptualize the nature of depressive cognitive disorders, some researchers (Zhu et al. 2025) have proposed that associated cognitive deficits in depression may be categorized into two types as reflected below. This method of categorization for cognitive sequelae of depression, described below, can be a useful tool in communicating functional experience of cognitive deficits to patients and their families when patterns on neuropsychological testing do not suggest an emerging neurodegenerative disorder.

"Cold" Cognition Entails nonemotional cognitive functions and processing deficits including the following:

- *Executive Function*: may include difficulty organizing, planning, making decisions, or problem-solving.
- *Memory*: may involve problems with short-term recall and/or learning of new information.
- *Attention*: includes the inability to focus or maintain concentration, particularly in busy environments or in the presence of competing information
- *Processing Speed*: entails slowing in mental processing of information and is often referred to as "brain fog" by patients presenting their concerns.

"Hot" Cognition Are emotionally laden biases where individuals prioritize negative information, such as:

- *Negative Bias*: Tendency to interpret neutral events negatively.
- *Rumination*: Persistent, uncontrollable focus on negative thoughts or failures.

This type of conceptual framework illustrates the array of cognitive distortions and deficits that contribute to a variety of functional cognitive symptoms present in 85%–94% of patients during depressive episodes and 39%–44% of the time during remissions in patients with major depressive disorder (MDD) (Ab Latiff et al. 2023). Thus, it is important for clinicians to avoid dismissing cognitive complaints in patients who may have co-existing depressive symptoms. Should a neurodegenerative disorder be ruled out following comprehensive evaluation, validation of cognitive symptoms experienced by patients with depression is essential to their understanding of the mechanisms underlying their cause. This requires explanation of how depressive symptoms such as the tendency toward negative self-evaluation (i.e., hot cognition) will adversely impact cognitive function (i.e., cold cognition) and/or the *perception* of cognitive performance errors.

Some clinicians may be guilty of dismissing the experience of cognitive difficulties in depressed patients. Within my own practice, multitudes of patients have described being informed by prior clinicians that "there is no dementia and you should just go see a psychiatrist." Although in theory if dementia has been ruled out and mental health factors are suspected a psychiatric referral may be indicated, statements such as these by a trusted healthcare provider can feel dismissive which only serves to perpetuate the stigma associated with mental health issues. Stigma regarding mental health is unfortunately persistent in society, as well as within healthcare settings (Knaak et al. 2017). However, perhaps the most direct impact to patient care is the consequent *self-stigma* which may lead to negative outcomes for depressed patients (Yanos et al. 2020). Such research suggests self-stigma, which indeed may be in reaction to communications within healthcare experiences, can result in the following:

- Reduced hopefulness for the future
- Lower self-esteem and self-worth
- Increased depressive and other psychiatric symptoms
- Increased difficulties with social relationships
- Reduced likelihood to persist/comply with mental health treatment
- Increased difficulties in work settings

Clearly the impact of stigma, whether it be societal or self-stigmatization, has wide ranging impact for the functional capacities in patients. For those individuals presenting to memory/dementia clinics, this may lead to further exacerbation of reported cognitive symptoms which serves only to increase likelihood of misdiagnosis, as described in Chap. 2, which will be accompanied by consequent outcome and treatment implications. In fact, research has revealed self-stigmatization results in significant barriers to treatment and more negative treatment outcomes for patients (Oexle et al. 2018).

Regarding the underlying mechanisms of memory dysfunction in depression, Dillon and Pizzagalli (2018) presented an overview, specifically noting that episodic memory (i.e., memory for everyday experiences and their details) is reduced in individuals with depression versus health controls. These researchers cite a replicated finding of reduced hippocampal volume (i.e., estimated at 4%–10% reduction bilaterally) in adults with depression versus healthy control, with a bias for negative material in depressed individuals reflecting research on the role of the amygdala (i.e., involved in emotionally charged memories) in this group as well. This research paper notes:

> Importantly, the relationship between depression and memory is bidirectional: depression affects memory, but memory problems likely exacerbate depression. A bias to repeatedly retrieve painful memories could clearly sustain a depressive episode, and failure to encode and consolidate positive memories could reinforce the anhedonia that (putatively) disrupted those processes in the first place. Thus, the role of memory in depression may need to be

reframed; although not commonly considered a core symptom of MDD, memory disruption may actually be critical to the negative sense of self that defines the disorder.

Depression primarily impacts episodic memory in later life as well as other cognitive areas such as processing speed and executive functions. Such cognitive change is hypothesized to occur through various biological mechanisms, including changes in brain structures like the hippocampus. Depression has been found to be associated with reduced volume in the hippocampus leading to increased forgetfulness and other cognitive impairments. Further, increased levels of stress hormones (i.e., glucocorticoids) related to depression have been hypothesized to result in brain tissue toxicity according to some researchers (Wilkins et al. 2009; Aziz and Steffens 2017).

Importantly, cognitive dysfunction has been identified as "under-evaluated" in the diagnosis of major depressive disorder (MDD), further serving as a mediator of psychosocial and functional outcomes (Zuckerman et al. 2018). This minimization of formal assessment for cognitive issues in MDD may lead patients to feelings of being unduly dismissed and unheard when reporting cognitive concerns associated with their depression. Formal evaluation of cognition in patients reporting decline or change may serve as an opportunity for clinicians to educate patients on the role of their depression in the expression of functional cognitive issues in their daily lives. Nevertheless, management of symptoms must be multifaceted including psychotherapy, pharmacological treatment, lifestyle strategies, as well as cognitive remediation.

Overall, the complexity of the relationship between depression and cognitive impairment continues to present challenges within clinical care, with depression and cognition mutually influencing each other (Yin et al. 2025). In this context, greater depressive symptoms were found to be associated with weaker memory at baseline and decline over time. The complex interplay of cognition and depression only further emphasizes the need for early and integrated treatment for depressive symptoms in patients presenting with cognitive complaints. It is important to acknowledge that such cognitive complaints may not only impact an individual's quality of life and functioning but may also increase recurrence risk for depression (Perini et al. 2019). Further, when depression presents in older adults, diagnoses of dementia and depression can have overlapping symptoms resulting in the reference to "pseudodementia" or depressive cognitive disorder (Sekhon et al. 2023).

As this issue is further confounded by the heterogeneity of dementia presentations and their prodromes, Tetsuka (2021) provided a comprehensive review of varying factors in distinguishing depressive cognitive disorders, noting the variance between dementia subtypes and their presentation. Despite factors that distinguish depressive cognitive disorders from neurodegenerative conditions, there are also commonalities which can result in challenges to diagnostic specificity. Thus, the importance of a fulsome evaluation of all factors that may contribute to presenting symptoms must be considered through considered analysis. The following table has been reproduced from Tetsuka's paper and provides a comprehensive review of distinguishing features of depressive cognitive disorder and various forms of dementia (Table 4.1):

Table 4.1 Depressive cognitive disorder versus dementia

	Depressive cognitive disorder	Dementia
Response and attitude toward functional decline	Attitudes that overestimate and pessimistically emphasize the decline in abilities	Lack of interest in or denial of diminished capacity, or an attitude of mending
Type of onset	Onset time can be identified on a weekly to monthly basis	Slow onset; onset time only identifiable in seasons or years
Variability and environmental reactivity	Diurnal variation that worsens in the morning; constancy even when the environment changes	Variations in attention and concentration suggest DLB; mood and motivation improve with positive environment and interpersonal interaction
Psychiatric symptoms	Feelings of sadness and remorse are present that are sometimes accompanied by feelings of hopelessness and thoughts of death, and sometimes accompanied by psychic delusions, delusions of guilt, and delusions of poverty; in rare cases, delusions of nihilism and immortality may accompany the symptoms	Lack of sadness and remorse, lethargy, and indifference (apathy) are the main symptoms. Emotional incontinence suggests VaD, sometimes accompanied by delusions of being robbed; repeated and specific visual hallucinations suggest DLB, mixed apathy and homophobic behavior suggest FTD
Movement symptoms	Not present	Parkinsonism and easy falling suggest DLB and VaD, and hemiplegia and dysarthria suggests VaD.
Sleep	Wakes up in the morning; going to sleep is an obstacle	Gradual rhythm disturbance, day and night reversal; REM sleep behavior disorder suggests DLB
Appetite, weight	Decreased appetite, sometimes increased appetite, with weight changes on a weekly to monthly basis	Slow weight loss: anorexia, overeating, and rapid weight gain suggest FTD
Simple cognitive function test Findings	Answers such as "I don't know" or "I don't remember" and slow thinking. Careless mistakes in continuous subtraction and reverse chanting; delayed playback is impaired, but reaffirmation is maintained; figure sketching and drawing will be maintained	Wrong answers and mending, and turning around signs. Thought laziness and walking away behavior suggest FTD; markedly delayed playback and impaired reification suggest AD; impairments in figure copying and drawing suggest AD and DLB
Morphological brain imaging	Normal or age-related changes, mild atrophy of the hippocampus, olfactory cortex, amygdala, and frontal lobe, and mild deep white matter ischemic changes	Atrophy of the hippocampus and olfactory cortex and parietal lobe suggest AD; medium to large infarcts, multiple infarcts, infarcts at strategic sites, and high levels of white matter lesions are indicative of VaD; severe frontal and temporal lobe atrophy suggest FTD

(continued)

Table 4.1 (continued)

	Depressive cognitive disorder	Dementia
Functional brain imaging (decreased cerebral blood flow and metabolism)	Normal or mildly impaired frontal lobe function	Functional decline in the posterior cingulate gyrus and anterior portion of the scapula suggests AD; functional decline in the occipital lobe suggests DLB; a high degree of frontal and temporal lobe dysfunction suggests FTD
Other functional imaging tests	Dopamine transporter and MIBG myocardial scintigraphy are normal	Decreased dopamine transporter uptake in the basal ganglia or decreased uptake on MIBG myocardial scintigraphy suggests DLB

Tetsuka (2021)
AD Alzheimer's disease, *VaD* vascular dementia, *DLB* dementia with Lewy bodies, *FTD* frontotemporal dementia

Given the entanglement of cognition and depressive disorders reflected above, and in addition to the considerations cited within the chart above, neuropsychological evaluation can play an important role in differential diagnosis of depressive cognitive disorder (DCD) versus emerging neurodegenerative disorders through identification of the pattern and consistency of cognitive deficits interpreted within the context of an individual's history.

In my own experience, this was well illustrated with a patient we will call Dan who was presenting to the clinic relatively young at age 58. Dan had been a police officer, retiring at age 50. He had several stressors in the years preceding his initial presentation to the clinic including a divorce and related financial issues. On longitudinal neuropsychological testing, Dan presented variability within cognitive domains (i.e., variability on multiple tests of memory and executive function including improved performances on more challenging tasks relative to easier tasks administered). He also presented considerable variability across the three neuropsychological assessments conducted, albeit with evidence of some decline in aspects of memory over time. Despite this variability, Dan's functional capacities were significantly declining over time which ultimately resulted in his admission to a care facility where he resided until his suicide in this facility approximately 10 years later. Dan had consented to brain autopsy at the time of his first presentation to clinic (i.e., years prior to his death). The autopsy revealed no neuropathological evidence of dementia. In this case, it was important to acknowledge Dan's functional decline and cognitive variability were likely the consequence of his marked depressive disorder, despite his trajectory of functional decline being suggestive of dementia. Certainly, such acknowledgment of depressive symptoms in clinical populations, where diagnosis of dementia is being considered, is critical to appropriate symptom management.

Although research suggests history of ignoring or dismissing cognitive symptoms in depression, scientists have more recently acknowledged the persistence of deficits in acute phases of depression as well as in long-term depressive symptoms (Singh et al. 2025). These researchers summarize findings on the course of cognitive symptoms in patients with major depressive disorder (MDD) in acute phases through longer term (i.e., 5 years following initial episode). This temporal evolution of cognitive deficits within various domains for individuals with major depressive disorder (MDD) was explored. The same researchers describe the course of MDD cognitive deficits, noted to be "moderate" to "strong," present in the acute phases of illness over time as reflected in a review of multiple studies in this area:

- *Psychomotor Speed*: Acute MDD demonstrates strong impairment in this area with impairment remaining in 40% of individuals with remitted MDD (i.e., when depressive symptoms have subsided).
- *Verbal Learning and Memory*: Noted to show strong deficits in acute MDD with moderate deficit in remitted MDD.
- *Visual Learning and Memory*: Also cited to show strong deficits in acute MDD, with the highest number of studies (i.e., 75%) reflecting deficits in this area within remitted MDD.
- *Attention and Working Memory*: Moderate deficits were found to be present in both acute and remitted MDD.
- *Executive Functions*: Moderate impairments were evident in this area for acute MDD, although no clear tendency was evident over studies with remitted MDD.
- *Verbal Fluency*: Moderate impairments were evident in this area for acute MDD, although no clear tendency was evident over studies with remitted MDD.

From the research of Singh and associates (2025), it is evident that moderate to strong deficits are present in acute phases of depression, along with persistence of cognitive deficits within all domains described years following initial depressive episode and variable deficits during remission. With evidence such as this, one can hypothesize the impact of persisting cognitive impairments when superimposed upon the cognitive change described earlier that is known to be associated with normal aging. Such persistence of cognitive impairment attributable to depression when combined with the course of cognitive change in normal aging can lead patients and in some cases clinicians to the conclusion that cognition is on a downward trend such as would be expected in neurodegenerative disorders. Once again, and recommended by these researchers, comprehensive evaluation is required to better understand such trajectories. Inherent in the analysis of findings from neuropsychological evaluations is comparison of cognitive performance with normative data for age, in essence to take the issue of age out of this equation when considering the degree of impairment present. Further, these researchers suggest longitudinal neuropsychological evaluation to enhance care planning but also to facilitate clinicians' understanding of cognition in depressive disorders.

With respect to symptom management, research suggests cognitive function also may be a predictor for response to treatment in depressed patients (Park et al. 2018), where cognitive dysfunction was noted to be a principal feature and determinant of health outcome in major depressive disorder (MDD). Such studies reflect cognitive dysfunction in depressed patients, particularly deficits in cognitive areas like executive function, attention, and memory, which can predict poorer response to standard antidepressant treatments. Studies also suggest early changes in cognition in areas such as emotional recognition might also signal future treatment success, highlighting cognition as a key factor for identifying patients' differing intervention needs (Thomas et al. 2009).

From this review, it is evident that more positive outcomes may be associated with earlier interventions and targeted cognitive rehabilitation. A multimodal approach to treatment of patients with cognitive symptoms in the context of depression is recommended, including pharmacological treatments, psychotherapy, lifestyle interventions, and cognitive rehabilitation. Although some of these approaches will be discussed in greater detail within subsequent chapters, with respect to cognitive rehabilitation approaches, both of the following may be considered:

- *Strategy-Based Training*: It involves teaching practical skills and techniques to work around cognitive difficulties. As mentioned previously, neuropsychological assessment can provide a "cognitive map" of strengths and weaknesses to guide such training (i.e., using cognitive strengths to compensate or support cognitive weaknesses).
- *Process-Based Training*: It involves practicing drills or exercises targeting specific areas of cognitive function in order to directly impact the underlying cognitive impairment.

Regardless of modality, cognitive rehabilitation is most often implemented by occupational therapists (OTs). Occupational therapists (OT), specifically neuro-OTs, are skilled in the translation of cognitive impairment regardless of etiology to functional adaptation. Consequently, having OTs working within the multidisciplinary teams of memory/dementia clinics seems to be an important step toward managing mental health and functional disorders within such settings.

Overall, incorporating a greater understanding of depressive cognitive disorders (DCDs) into comprehensive investigations of neurodegenerative disorders is essential in clinical practice. Supporting patients with clinically significant depressive symptoms is an important step toward comprehensive care. Unfortunately, the stigma associated with mental health issues remain, both societally and personally. This self-stigmatization is often demonstrated in the portion of patients that I see who appear almost disappointed when informed their cognitive presentation may be associated with depression or other mental health issues rather than dementia. This disappointment when informed they do not have patterns consistent with dementia seems counterintuitive from a clinical perspective, given that mental health issues may potentially be effectively treated and resolved, while no such reversal exists for dementia. Clinicians' approach to such patients must not lead with judgment or dismissal, despite the apparent distortion of logic in the patient's (or their family's) seeming preference for a dementia diagnosis over a reversible condition such as depression. Approaching patients with supportive psychoeducation regarding the complex relationship of depression, cognition and the aging brain are essential for receptivity to future mental health treatment and outcomes. This form of psychoeducation is often facilitated by comprehensive cognitive evaluation to inform patients of their measured cognitive skills relative to age appropriate normative data, which may serve to mitigate misinterpretation of cognitive capabilities as well as any exisiting self-stigmatization.

References

Ab Latiff HZ, Ariaratnam S, Shuib N, Isa MR. Cognitive decline and its associated factors in patients with major depressive disorder. Healthcare (Basel). 2023;11(7):950. https://doi.org/10.3390/healthcare11070950.

Aziz R, Steffens D. Overlay of late-life depression and cognitive impairment. Focus (Am Psychiatr Publ). 2017;15(1):35–41.

Brodaty H, Connors MH. Pseudodementia, pseudo-pseudodementia, and pseudodepression. Alzheimers Dement (Amst). 2020;12(1):e12027.

Devita M, De Salvo R, Ravelli A, De Rui M, Coin A, Sergi G, Mapelli D. Recognizing depression in the elderly: practical guidance and challenges for clinical management. Neuropsychiatr Dis Treat. 2022;18:2867–80.

Dillon DG, Pizzagalli DA. Mechanisms of memory disruption in depression. Trends Neurosci. 2018;41(3):137–49. https://doi.org/10.1016/j.tins.2017.12.006.

Donovan NJ, Locascio JJ, Marshall GA, Gatchel J, Hanseeuw BJ, Rentz DM, Johnson KA, Sperling RA. Harvard Aging Brain Study. Longitudinal association of amyloid beta and anxious-depressive symptoms in cognitively normal older adults. Am J Psychiatry. 2018;175(6):530–7.

Kiloh LG. Pseudo-dementia. Acta Psychiatr Scand. 1961;37(4):336–51.

Knaak S, Mantler E, Szeto A. Mental illness-related stigma in healthcare: barriers to access and care and evidence-based solutions. Healthc Manage Forum. 2017;30(2):111–6.

Kriesche D, Woll CFJ, Tschentscher N, Engel RR, Karch S. Neurocognitive deficits in depression: a systematic review of cognitive impairment in the acute and remitted state. Eur Arch Psychiatry Clin Neurosci. 2023;273(5):1105–28.

Livingston G, Huntley J, Sommerlad A, Ames D, Ballard C, Banerjee S, Brayne C, Burns A, Cohen-Mansfield J, Cooper C, Costafreda SG, Dias A, Fox N, Gitlin LN, Howard R, Kales HC, Kivimäki M, Larson EB, Ogunniyi A, Orgeta V, Ritchie K, Rockwood K, Sampson EL, Samus Q, Schneider LS, Selbæk G, Teri L, Mukadam N. Dementia prevention, intervention, and care: 2020 report of the Lancet Commission. Lancet. 2020;396(10248):413–46.

Ma L. Depression, anxiety, and apathy in mild cognitive impairment: current perspectives. Front Aging Neurosci. 2020;12:9.

McWhirter L, Ritchie C, Stone J, Carson A. Functional cognitive disorders: a systematic review. Lancet Psychiatry. 2020;7(2):191–207.

Mouta S, Fonseca Vaz I, Pires M, Ramos S, Figueiredo D. What do we know about pseudodementia? Gen Psychiatr. 2023;36(4):e100939.

Oexle N, Müller M, Kawohl W, Xu Z, Viering S, Wyss C, Vetter S, Rüsch N. Self-stigma as a barrier to recovery: a longitudinal study. Eur Arch Psychiatry Clin Neurosci. 2018;268(2):209–12.

Park C, Pan Z, Brietzke E, Subramaniapillai M, Rosenblat JD, Zuckerman H, Lee Y, Fus D, McIntyre RS. Predicting antidepressant response using early changes in cognition: a systematic review. Behav Brain Res. 2018;353:154–60.

Paula JJ, Miranda DM, Nicolato R, Moraes EN, Bicalho MA, Malloy-Diniz LF. Verbal learning on depressive pseudodementia: accentuate impairment of free recall, moderate on learning processes, and spared short-term and recognition memory. Arq Neuropsiquiatr. 2013;71(9A):596–9.

Perini G, Cotta Ramusino M, Sinforiani E, Bernini S, Petrachi R, Costa A. Cognitive impairment in depression: recent advances and novel treatments. Neuropsychiatr Dis Treat. 2019;15:1249–58.

Rhee TG, Shim SR, Manning KJ, Tennen HA, Kaster TS, d'Andrea G, Forester BP, Nierenberg AA, McIntyre RS, Steffens DC. Neuropsychological assessments of cognitive impairment in major depressive disorder: a systematic review and meta-analysis with meta-regression. Psychother Psychosom. 2024;93(1):8–23.

Sekhon S, Patel J, Sapra A. Late-life depression. In: StatPearls [Internet]. Treasure Island (FL): StatPearls Publishing; 2023. PMID: 31855351.

Semkovska M, Quinlivan L, O'Grady T, Johnson R, Collins A, O'Connor J, Knittle H, Ahern E, Gload T. Cognitive function following a major depressive episode: a systematic review and meta-analysis. Lancet Psychiatry. 2019;6(10):851–61.

Singh OP, Avinash PR, Gautam A, Prasad A, Jagawat T. Clinical practice guidelines on the assessment and management of cognitive impairment in major depressive disorder. Indian J Psychiatry. 2025;67(1):98–105.

Smith GS, Kuwabara H, Nandi A, Gould NF, Nassery N, Savonenko A, Joo JH, Kraut M, Brasic J, Holt DP, Hall AW, Mathews WB, Dannals RF, Avramopoulos D, Workman CI. Molecular imaging of beta-amyloid deposition in late-life depression. Neurobiol Aging. 2021;101:85–93.

Tetsuka S. Depression and dementia in older adults: a neuropsychological review. Aging Dis. 2021;12(8):1920–34.

Thomas AJ, Gallagher P, Robinson LJ, Porter RJ, Young AH, Ferrier IN, O'Brien JT. A comparison of neurocognitive impairment in younger and older adults with major depression. Psychol Med. 2009;39(5):725–33.

Wilkins CH, Mathews J, Sheline YI. Late life depression with cognitive impairment: evaluation and treatment. Clin Interv Aging. 2009;4:51–7. Epub May 14. PMID: 19503765; PMCID: PMC2685224.

Yanos PT, DeLuca JS, Roe D, Lysaker PH. The impact of illness identity on recovery from severe mental illness: a review of the evidence. Psychiatry Res. 2020;288:112950.

Yin J, John A, Cadar D. Bidirectional relation of depressive symptoms and cognitive function over 16 years follow-up. Alzheimers Dement. 2025;20(Suppl 8):e095408.

Zhu N, Zhang Q, Huang J, Tong J, Gong HF, Zhu MH, Lu W, Zhang J, Sun XR. Using the THINC-integrated tool to compare the characteristics of cognitive dysfunction in patients with unipolar and bipolar depression. World J Psychiatry. 2025;15(3):99408.

Zuckerman H, Pan Z, Park C, Brietzke E, Musial N, Shariq AS, Iacobucci M, Yim SJ, Lui Leanna MW, Rong C, McIntyre RS. Recognition and treatment of cognitive dysfunction in major depressive disorder. Front Psych. 2018;9:655.

5 Fear of Forgetfulness: Anxiety and Cognition

One of the primary inspirations for the writing of this book was my experience navigating the clinical challenges associated with patients presenting cognitive complaints in the context of significant anxiety symptoms. This "fear of forgetfulness" has been a prominent part of my own practice, with over 60% of the patients I see for neuropsychological evaluation in the Clinic for Alzheimer's Disease and Related Disorders (CARD) presenting with measurable levels of anxiety and stress.

Research has suggested that high levels of fear for memory loss may lead to increased incidence of memory failures and sleep disturbance (Farina et al. 2023). These researchers also found that individuals with forgetfulness-related fear avoidance (i.e., behavioral pattern related to avoiding situations that may trigger fear) experienced measurable impairments in memory, including decreased verbal recall often present in the prodromes of dementia, as well as reduced social functioning and quality of life. Thus, the clinical significance of fear or anxiety cannot be underestimated within the scope of diagnostics for neurodegenerative conditions particularly given the symptom overlap for anxiety-based conditions with dementia and its prodromes.

This concept has been well validated within emerging literature speaking to the bidirectional relationship between anxiety and cognition (Jain et al. 2023), with these researchers noting:

> "clinicians should be sensitive both to potential consequent anxiety when patients present with cognitive impairment, and to potential incipient MCI when the presenting complaint is anxiety. Managing stress may help mitigate adverse outcomes." (Jain et al. 2023).

So how exactly does anxiety impact cognition? Often when speaking with patients about the role of anxiety in cognitive performance and perception of performance, I must spend an extended period explaining this complex relationship in a way that is palatable and understandable. Although this may seem an unnecessary use of precious clinical time, it is a critical component of clinical care given that many patients I see have significant distortions in perception regarding the degree

S. Hayden, *Fear of Forgetfulness*, https://doi.org/10.1007/978-3-032-21818-6_5

and nature of their cognitive performance, because of their anxiety. The issue of self-stigmatization applies to anxiety as it does to depression discussed within the last chapter.

Aspects of cognition most directly impacted by anxiety are often within the arena of attention and concentration, which in turn can impact memory on a variable basis. Anxiety impacts attention and concentration by acting as a maladaptive behavior through constantly shifting focus toward potential threats rather than the task at hand. While normally our brain can choose what to focus on, anxiety hijacks this system making it difficult to concentrate on work, conversations, or studies because the brain is busy scanning for danger. Unfortunately, this maladaptive use of the fight-flight response then competes for attention in an individual. Subsequently, the patient will find themselves with only *sporadic* recall of the material they had been presented with previously. Often with anxious patients in my practice, I will explain how our brains normally screen out such competing "noise" but with the background of anxious thoughts competing for attention, the encoding of new material is inefficient. It is for this reason that many of these patients will acknowledge the variable nature of their forgetfulness on a functional basis. Providing explanation on how their memory system (i.e., encoding or intake of information) is adversely impacted by anxious thoughts can provide a framework for patients to understand forgetfulness that may not be attributable to dementia. Thus, within clinical interviews, health providers should not only ask about occurrence of short-term memory loss but also if such occurrences are variable or consistent in nature (Hill et al. 2019; Kwak et al. 2017).

As with depression, hypothesized mechanisms underlying the relationship of anxiety with cognition are bidirectional in nature, with resulting cyclic anxiety and cognitive dysfunction (Fig. 5.1).

More specifically, with anxiety there is impact on working memory/attentional processes, as anxiety can consume cognitive resources related to thoughts of worry or perceived threat. This disruption with the encoding of information for anxious individuals will, in turn, adversely impact subsequent recall of such information, which often may lead to triggering of increased anxiety. Further, the "fight-or-flight" response associated with anxious thoughts and perceived threats results not only in elevated cortisol levels and sleep disruption, but can further tax cognitive resources, impacting attention and memory, often then re-triggering anxiety as illustrated within the graphic above.

In my experience, it is helpful to have *measurable* evidence of anxiety on psychological measurement in these patients by using validated psychological tools which can subsequently facilitate clinician to patient communications as described earlier. It is for this reason, that once again neuropsychological assessment can provide a useful tool for communication in this regard. I commonly go through the results of psychological testing, including the findings evident on measurement of anxiety, which I then follow with an explanation of how that level of anxiety may adversely impact attention/concentration and then subsequently memory. For the purpose of demonstration, such communication by a clinician to patient might theoretically go something like this:

Fig. 5.1 Cycle of anxiety and forgetfulness

> So, responses on the mood measurement of your assessment reflects anxiety at the time of testing to be in the 98th percentile, meaning you are above 98 percent of the population for that form of anxiety. With anxiety of this degree, you might imagine that cognition will no doubt be impacted. For example, anxious thoughts can consume your attention, which in turn can compete or distract you from incoming information resulting in limited encoding of such information. If the information is not 'encoded' or entered into the memory system, it most certainly won't be remembered later. This is how anxious thoughts can impact, even mimic, memory issues seen in dementia. However, the pattern on neuropsychological testing shows inconsistent issues in attention which in turn seem to impact your memory on a variable basis. This is why I suspect the day-to-day difficulties in memory you have noticed may be related to anxiety. It is important to know that I certainly believe these memory issues are frustrating and disruptive to you. However, I would say though that the cause of your forgetfulness appears to be related to mental health, or these anxious thoughts. The good news is there are ways that this underlying cause of your forgetfulness can be treated.

This explanation then allows for and naturally leads into the next step of proposing my recommended treatment plan to the patient and their family (if present), based upon the neuropsychological assessment. In my experience, such patients are more receptive to considering mental health treatment with this approach to communication, even when they have had prior perceptions and/or convictions that their functional issues are related to an emerging neurodegenerative condition. This is certainly a more supportive and palatable means of clinical communication than the tendency to dismissively state "there is no sign of dementia so you should seek psychiatric treatment."

Consideration of communications related to mental health and cognition requires ongoing improvements on the clinical front lines given research that suggests over 32% of patients feel their cognitive concerns were dismissed by their medical provider (Hill et al. 2023). These researchers note that communication is critical for early detection of mild cognitive impairment (MCI) despite an estimate that upward of 90% of individuals with MCI go undetected, and under 20% of older adults discuss cognitive concerns with their healthcare providers. Stigmatization, which is not only directed at mental health as described earlier, may also relate to dementia and further complicates the diagnostic issues further. Notably, a journal editorial in Lancet (2024) cited a finding from Alzheimer's Disease International (ADI), 2024 World Alzheimer Report wherein 80% of the general public were found to believe dementia is a natural part of aging in 2024, which represents an increase in this belief relative to 2019. Furthermore, two-thirds of healthcare providers "believe dementia is a normal part of aging" reflecting slight increase in this bias since 2019 (i.e., from 62% in 2019 to 65% in 2024). Obviously, such evidence of increased healthcare biases as well as societal stigmatization will translate to marked challenges in the clinical care and management of mental health conditions as well as dementia and its prodromes. Such perceptual distortions may also serve to exacerbate fearfulness and anxiety syndromes pertaining to memory loss.

Regardless, there is complexity in the presentation and diagnostic specificity of patients presenting with cognitive concerns and co-existing anxiety, as anxiety can also be a common symptom in the prodromal and early stages of neurodegenerative disease (Chen et al. 2018; Gulpers et al. 2022). The change or loss of cognitive function related to an emerging dementia can manifest in worry about such losses as well as anticipatory anxiety about "what if I forget." In early-stage dementia, patients often recognize not only the cognitive change experienced but also experience the fear and stress related to concerns for future decline in such conditions. Furthermore, the neurological changes associated with many neurodegenerative conditions can, in of themselves, impact emotional regulation which can include directly eliciting anxiety-based responses. These issues may worsen symptoms for those even in the early stages of neurodegenerative conditions.

For both patients presenting with cognitive concerns primarily related to anxiety and for those with secondary anxiety related to cognitive change in early dementia, the impact of anxiety on attention/working memory can be dependent upon a task's complexity. With low to medium complexity cognitive tasks, anxiety significantly impairs cognitive performance by allowing more resources for threat processing as reflected within the illustration provided earlier. With cognitive tasks that are highly complex, fewer resources are available for threats, so working memory is less impaired by anxiety (Robinson et al. 2013). In this way, anxiety may result in errors on seemingly simpler tasks which, in turn, can lead to further distortions of perception and fearful/anxiety-based responses. I have witnessed this issue exemplified in patients who state "I seem to forget even simple things but at work I am doing ok," this becomes evidence in their anxious brains that a dementia is emerging. Again, in-depth clinical interview skills (i.e., exploring patient's perceptions of cognitive performance) by health providers may be one step in reducing such misperceptions.

Of course, it is important to note there is wide diversity in conditions that fall under the category of anxiety, likely with differing impact upon cognition although specificity of this is limited within current research. However, there is a specific label for anxiety related to cognitive dysfunction called athazagoraphobia, which will be discussed below.

Athazagoraphobia

One maladaptive form of anxiety related to cognition dysfunction entails the intense and irrational fear of forgetting or of losing one's memory, which is one type of *athazagoraphobia*. Symptoms of athazagoraphobia can include the following:

- Intense anxiety or panic attacks at the thought of forgetting
- Difficulty concentrating
- Impairment in processing speed
- Constantly seeking reassurance
- Social isolation
- Physical symptoms like a racing heart or sweating

Of note, there are multiple factors associated with athazagoraphobia, although causality is often complex. One such linkage to athazagoraphobia is certainly an individual's worry about contracting dementia, often with specific concerns for developing Alzheimer's disease as this is the condition most prominent in mass culture. Such fears may be particularly heightened in individuals with family history of Alzheimer's disease/dementia and/or in those with increased exposure to the disease, such as caregivers of dementia patients (Suhr and Kinkela 2007). Further, as mentioned previously, the anticipatory worry surrounding dementia biomarker communications may be contributory for this condition as well.

The need for comprehensive review of trauma and mental health history within dementia evaluations is further reflected in evidence for causality of athazagoraphobia. Traumatic experiences, particularly those involving abandonment, neglect, or loss, are considered to be significant contributing factors for developing athazagoraphobia. As such trauma may occur many years, and in some instances decades, prior to the formal assessment for dementia, review of such factors is typically excluded. Regardless, the etiological connection of athazagoraphobia to trauma includes the following issues which may be explored when assessing history of patients in memory/dementia clinics:

- *Childhood Trauma*: This may include experiences such as being left alone for extended periods, emotional neglect, or inconsistent caregiving during childhood which may instill a deep-seated fear of insignificance and being forgotten later in life or forgetting oneself. Physical, psychological, and sexual abuse in childhood are also risk factors for fear of memory loss later in life.
- *Loss and Separation*: The sudden loss of a loved one or repeatedly being left behind in relationships can intensify anxiety about separation and being "erased from memory" which can lead to preoccupation with memory loss. The trauma of such losses can also manifest as athazagoraphobia.
- *Witnessing Memory Loss*: Exposure to loved ones suffering from memory-related conditions like Alzheimer's or dementia can be a traumatic experience, triggering fears of forgetting or being forgotten by those individuals which again may result in symptoms of athazagoraphobia.

The contributing mechanisms of impact for such traumatic events on fear of forgetfulness remains somewhat unclear, although the trauma may lead to issues such as hypervigilance (i.e., state of constant alertness to cognitive errors), low self-esteem and self-doubt (i.e., leading to "second guessing" of cognitive performance), and in some cases dissociative symptoms (i.e., there may be loss of memories during periods of trauma). Throughout my career, I have had patients with significant trauma history often in their childhood that has resulted in dissociative symptoms in the form of memory gaps for large sections of their past life (i.e., "I can't remember if I went to high school"). Once again, if dementia can be ruled out, it is important to provide education regarding the relationship of trauma to anxiety and memory formation in order to ease the patient's fear related to cognitive decline.

The role of such traumatic events or experiences in the development of anxiety for forgetfulness, such as athazagoraphobia, is hypothesized to involve two major mechanisms of connection:

- *Reshaping Cognitive Patterns*: Trauma can alter an individual's core beliefs about safety and self-worth, making them feel incompetent, invisible, or that the world is an unsafe place. Athazagoraphobia can be an expression of these deep-rooted insecurities and fears resulting from past experiences, which certainly can become more entrenched or exacerbated as individuals face the challenges and inevitable losses associated with normal aging, including normal cognitive change cited earlier.
- *Overlap with Other Trauma-Related Conditions*: Athazagoraphobia often overlaps with symptoms of other anxiety disorders including posttraumatic stress disorder (PTSD). Such conditions result in symptoms including as avoidance behaviors and hyperarousal (a heightened sense of awareness to potential danger) which can trigger or exacerbate the presentation of athazagoraphobia.

Overall, athazagoraphobia is frequently understood as a potential long-term consequence of past trauma, particularly experiences that involve a real or perceived threat to one's sense of connection, safety, and personal significance. Acknowledging and addressing this trauma history is often an essential part of the treatment for athazagoraphobia and thus must be considered with patients expressing marked fear of forgetfulness.

Although discussing a patient's history of trauma, such as suggested above, is not typically a component inherent within dementia investigations, the association of trauma with athazagoraphobia as well as risk for cognitive decline (Severs et al. 2023) suggests clinicians should be incorporating such reviews within their standard practice with this population. Within my own clinical interview with patients in our Clinic, it has become standard for my practice to review history of all mental health issues including trauma, for both historically remote and recent trauma events.

In one such instance, despite no reference to trauma in her medical history, a 72-year-old patient, we will call Bernice, indicated within neuropsychological interview that she had sexual and physical abuse within her family of origin during childhood, along with physical and psychological abuse within subsequent adult relationships. Such factors had not been evident in prior medical documentation on Bernice, as it was not considered the focus of care in her presentation of memory concerns. Although Bernice had been guarded in interview, stating she did not trust psychologists, she had inadvertently revealed the abuse when discussing her educational history (i.e., "I missed a lot of school"). This then gave opportunity for my questions about why she had missed school, leading Bernice to reveal her history of trauma and abuse. Although she would not discuss the nature of abuse in detail, her acknowledgment of these past experiences served to inform subsequent clinical care. The neuropsychological evaluation revealed Bernice's perceived memory deficits were likely related to the variable performance on attentional tasks as well as indicators of mental health issues on psychological testing (i.e., anxiety and post-traumatic stress disorder). With considered and open discussion of the significance of her past trauma upon her aging brain, Bernice acknowledge the need for further mental health supports to manage her symptoms.

Within research, history of lifetime trauma has been associated with cognitive impairment in mid-life and beyond (Lynch and Lachman 2020). Research such as this indicates trauma exposure may be a risk factor for cognitive decline in adulthood. Consequently, although not typically considered within the context of assessment for neurodegenerative disorders, *trauma-informed care* is critical to understanding the context within which a patient may experience cognitive concerns. Trauma-informed care in its most rudimentary form is a framework grounded in the understanding of trauma and responsiveness to the impact of trauma in patient care. This approach to care is particularly important given estimates suggesting 70% of older adults have experienced at least one psychologically traumatic event in their lifetime (Couzner et al. 2022).

Other researchers have found that adverse childhood experiences (ACEs) may raise risk for dementia (Abouelmagd et al. 2024). Although the underlying mechanism for connection between psychological trauma and neurodegenerative conditions is not fully understood, the association suggests exploration of trauma should be incorporated within dementia assessment. Given the high prevalence of trauma history in aging populations and trauma as a possible risk factor for cognitive decline, queries regarding possible trauma events must be incorporated into their clinical care including service focused upon investigation of dementia and its prodromes to ensure full understanding of contributing factors.

Stress, Anxiety, and Cognition

Long-term stress also appears to be linked to or may represent a possible trigger for athazagoraphobia as well as anxiety disorders in general. High perceived stress and its relationship with perceived threat of Alzheimer's disease was explored by Ostergren et al. (2017) who note high perceived stress is correlated with increased perceived threat of Alzheimer's disease. Interestingly, they also found significant associations between personal beliefs about risk factors for Alzheimer's disease and perceived threat for this condition, with stronger endorsement of the importance of stress or genetics as risk factors being associated with greater perceived threat for Alzheimer's disease. These researchers suggest the "association between these risk factor beliefs and heightened perceived AD threat may also indicate respondents who strongly endorse genetics or stress as risk factors may feel they have less control over the disease and view it as less preventable." This supports prior findings that the experience of lower levels of control beliefs are associated with higher levels of anxiety, which in turn adversely impacts cognition, specifically via task interference which impedes episodic memory (Lachman and Agrigoroaei 2012). Notably, the public perception and understanding of risk for dementia often overestimates the role of genetics in these conditions. Within the context of this theoretical framework, this overestimation of genetic risk (i.e., "my mother had Alzheimer's disease so I am going to get it") may be contributing to heightened anxiety or perceived threat of Alzheimer's disease.

In a paper on the role of stress in Alzheimer's disease, Justice (2018) affirmed that neurodegenerative diseases naturally lead to loss of cognitive function over time which is certainly stressful. However, this researcher also describes how this experience of distress then disrupts the brain pathways that mediate the stress response. In turn, this stress response exacerbates disease symptoms and may drive disease progression, including exacerbation of neuropsychiatric symptoms in an ongoing cycle with resulting cognitive decline.

Franks et al. (2021) expanded on such concepts within a meta-analysis of research examining stress and risk for MCI and dementia. Across over 1600 studies screened, higher perceived stress was significantly associated with an increased risk of mild cognitive impairment (MCI) as well as with increased risk for dementia of all causes. Notably, being exposed to two or more stressful life events (i.e., versus none) was associated with an increased risk of dementia, while one or more stressful life events did not increase risk for dementia.

Overall, such literature along with my own experience in clinical practice supports a significant role for stress and anxiety in the risk for MCI and dementia as well as the exacerbation of symptoms in these conditions. This most certainly reflects increasing need for development of policies and supports specific to stress reduction as a modifiable risk factor in dementia (Graham-Engeland et al. 2025) and its prodromes.

On the clinical front, the issue of stress within the clinical presentation of cognitive concerns may again be addressed with a comprehensive approach to interview and formalized assessment of such factors. In my own practice, this includes specific questions regarding past and present perceived stressors that a patient may be experiencing. However, it is not uncommon for patients to underreport or minimize such stressors when presenting to investigation for memory disorders as the individual may not connect such stressors with functional cognitive performance. Thus, exploration of potential stressors may also occur within collateral interview (i.e., with family or close friend) that I conduct separately while a patient is undergoing their cognitive testing in another room. Characterizing the nature and timeline of a patient's stressors may provide important clues to their clinical cognitive presentation and its reported course. I have commonly found patients who describe onset of cognitive symptoms within the same timeframe as they later cite the onset or presence of stressors, and at times without awareness of the coinciding nature of these timelines.

Obsessive-Compulsive Disorder and Cognitive Disorders

Fear of forgetfulness can also manifest clinically in the form of an obsessive concern about memory. In some cases, this may be linked to obsessive-compulsive tendencies where individuals constantly seek reassurance or engage in rituals to "remember" things correctly further exacerbating underlying anxiety/fear. In fact, emerging research (Ramos and Salgado 2023) suggests a significant link between obsessive-compulsive disorder (OCD) and risk for dementia. This study noted an estimated four times higher risk for developing Alzheimer's disease and vascular dementia in individuals with OCD. Notably, the dementia diagnoses in these individuals with OCD appears to occur about 6 years earlier (i.e., around age 70.5) compared to those without OCD (i.e., around age 76.7). Further, early-onset dementia (i.e., before 65) was reportedly more common in people with OCD than those without this condition. It is also important to recognize that other research suggests neuropsychological impairment may be associated with OCD *distinct* from dementia diagnoses, with significant impairment evident in executive function and nonverbal memory reflecting frontal striatal involvement for individuals diagnosed solely with OCD (Suhas and Rao 2019).

Despite establishment of the connection between OCD and risk for dementia in current research, the reason for this connection is less clear. Nevertheless, both individuals diagnosed solely with OCD and patients with some forms of dementia can show cognitive deficits in executive function and memory which suggests overlapping neural systems with proposed shared biological brain pathways. One proposed neural pathway involves *glutamate* which is the brain's main excitatory neurotransmitter, acting somewhat like the accelerator for neurons to fire. Glutamate is thought to be crucial for learning and memory, with glutamatergic dysfunction leading to excitotoxicity or receptor problems linked to both OCD and dementia (Karthik et al. 2020). For both OCD and dementia, this seems to reflect having the accelerator on for too long causing glutamate to build up and becoming toxic or causing neural circuits to misfire.

When evaluated formally through psychological measurement in clinical setting, it is not uncommon for my patients to endorse markedly elevated anxiety and/or obsessive-compulsive traits when presenting with concerns for cognitive decline. Although anxiety and OCD are distinct conditions, symptoms for both can include the following in the context of fear of forgetfulness:

- Intense anxiety or panic attacks at the thought of forgetting
- Difficulty concentrating, which in turn may result in sporadic forgetfulness
- Seeking of reassurance or checking behaviors related to memory
- Social isolation due to anticipatory anxiety about forgetfulness
- Physical symptoms like a racing heart or sweating when faced with cognitive errors

The symptoms of OCD, as with anxiety, will clearly interfere with the encoding or intake of information which will in turn adversely impact recall of such material which then further exacerbates the fear/anxiety for memory loss. Research (Suhas and Rao 2019) supports the concept that OCD alone can have a direct, detrimental impact on memory and cognitive function, including impact within the following cognitive domains:

- *Executive Functioning*: Individuals with OCD are commonly found to have deficits in planning, decision-making, and organizational skills on a functional basis.
- *Cognitive Inflexibility (Set-Shifting)*: OCD patients may have difficulty shifting attention or behaviors, contributing to their rigid, ritualistic actions.
- *Response Inhibition*: OCD entails a reduced ability to suppress prepotent motor responses, which may contribute to their inability to stop unwanted behaviors or thoughts (i.e., obsessions and compulsions).
- *Memory Impairment*: Deficits in nonverbal (visuospatial) memory are reported in OCD patients, which seems consistent in the support of some checking behaviors.
- *Processing Speed*: Slowed information processing may be present in OCD patients, which could also underlie deficiencies in other cognitive areas.
- *Memory Confidence*: Individuals with OCD often report lower confidence in their memory, rather than a lack of memory capacity as reflected on cognitive testing.

The pattern of forgetfulness in OCD is complicated further by the "anxiety-forgetfulness cycle." As mentioned, forgetfulness caused by anxiety often becomes a new source of worry, which for some can exacerbate their obsessive-compulsive tendencies. In turn, these OCD symptoms lead to more forgetfulness, perpetuating a difficult cycle with the illusion of cognitive decline. Further, anxious thoughts can

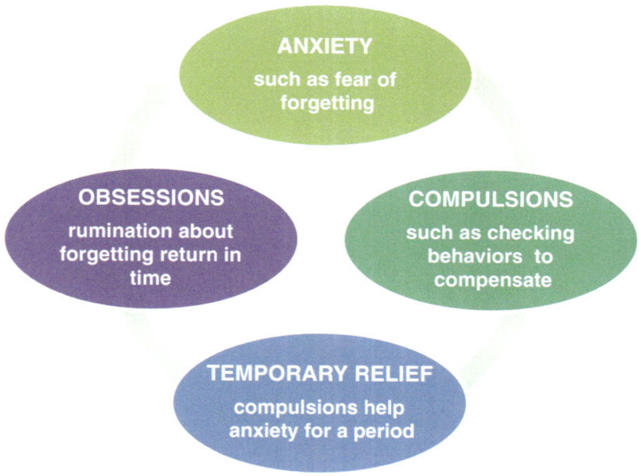

Fig. 5.2 OCD cycle and cognition

create a negative view of one's own memory, making individuals believe they are performing worse than they are, which is a cognitive distortion (i.e., inaccurate perception) evident in many anxiety disorders including OCD.

This negative cycle can be further sustained by a fear-avoidance model, where perceived memory changes lead to fear, which results in social withdrawal and avoidance behaviors, further impacting daily life and self-esteem. Fear avoidance is a core feature of OCD with the avoidance behaviors often reinforcing OCD cycles, depicted below, as they do not learn irrationality of their fear driven OCD behaviors (Fig. 5.2).

Posttraumatic Stress Disorder (PTSD) and Cognition

Another anxiety-based mental health condition with associations to both anxiety and trauma is posttraumatic stress disorder (PTSD). PTSD has also been found to significantly increase the risk for developing dementia (Prieto et al. 2023). Some studies reflect an up to a twofold higher risk for dementia in people with diagnosis of PTSD. This is potentially due to shared biological pathways involving inflammation, sleep disruption, amyloid buildup, and brain structural changes, while also being a bidirectional link where dementia can worsen PTSD symptoms (Desmarais et al. 2020). The severity of PTSD also appears to have a dose-dependent relationship with dementia, as higher PTSD severity is associated with more elevated dementia risk, including Alzheimer's disease albeit with some evidence pointing to stronger links with frontotemporal dementia (FTD) and vascular dementia (VaD).

Potential mechanisms linking PTSD and dementia risk requires further research, although the mechanisms below have been proposed by some scientists (Mohlenhoff et al. 2017; de Araujo et al. 2023). Such hypotheses include the following:

- *Inflammation*: PTSD can elevate inflammatory markers (i.e., cytokines) that can damage neurons resulting in impaired learning and memory.
- *Sleep Disturbances*: Poor sleep, which is common in PTSD, hinders brain repair and promotes amyloid protein buildup further contributing to inflammation. Also, poor sleep resulting in persistent fatigue will adversely impact attention, and as a consequence memory due to poor encoding of information to be recalled.
- *Amyloid and Tau*: Sleep disturbance linked to PTSD may accelerate the accumulation of amyloid-beta proteins, though some research shows PTSD might cause cognitive issues without elevated Alzheimer's biomarkers like amyloid or tau.
- *Brain Structure Changes*: Neuroimaging studies link PTSD to changes in brain structures crucial for memory and executive function. This includes studies associating reduced volume in the hippocampus and the prefrontal cortex with PTSD.
- *Co-occurring Conditions*: High rates of depression, traumatic brain injury, diabetes, and cardiovascular issues in PTSD patients also independently raise dementia risk in this population.

Regardless of mechanism for causality, the connection between dementia and PTSD is not one way as with conditions discussed previously. Dementia can also increase the risk of delayed-onset PTSD or worsen existing PTSD in trauma survivors, indicating a complex interplay of symptoms across all these conditions (Desmarais et al. 2020). This appears further complicated by sex-based differences with some research indicating a potentially stronger association in women with PTSD in middle-aged women and cognitive decline (Roberts et al. 2022). Although PTSD does not consistently predict dementia onset, it certainly appears to represent a risk factor. Thus, addressing PTSD symptoms like sleep disturbance through trauma-informed care can improve well-being and potentially reduce dementia risk.

Perfectionism and Cognition

One of the biggest challenges to stress adaptation and emotional balance appears to be related to maladaptive perfectionistic personality traits. Perfectionism refers to a personality trait focused on the unhealthy, relentless "striving to be flawless" which can create the unbearable burden of unachievable goals. Commonly, the consequence of this burden is a persistent feeling of inadequacy or the imposter syndrome which can be experienced even by individuals who appear outwardly well accomplished. Perfectionism is linked to a multitude of maladaptive behavior patterns, including procrastination (i.e., which entails the unnecessary delay of an important task, despite knowing that this delay will likely lead to negative consequences). These behaviors are an additional barrier to managing stress effectively and prevent you from seeking interventions that lower your risk for related mood disorders.

In a systematic review of 118 articles about maladaptive perfectionism and potential treatments, Patterson et al. (2021) suggest the foundation for a perfectionistic attitude is unrealistic or unattainable standards and excessive self-criticism. Clinical studies also suggest perfectionism may be linked to various forms of cognitive impairment, primarily through the depletion of mental resources and the use of maladaptive processing strategies as in anxiety. This relationship is often characterized by a discrepancy where individuals *feel* cognitively impaired even when objective testing shows standard performance (Renaud and Lacroix 2025).

Certainly, I have seen many such patients in my practice who provide detailed recollection of their various perceived cognitive errors within the context of marked intolerance for such cognitive errors, even when they are performing within expectation for normal aging. However, identifying perfectionism within clinical settings can be a challenge as often such individuals have excelled on a functional basis, including attaining high levels of occupational status. Societal expectations of performance support perfectionism, as this trait is typically associated with high levels of performance. When such traits cross into the territory of maladaptation, there is little differentiation from consequence of anxiety issues and perfectionism upon cognition. High levels of perfectionism may be associated with poorer self-reported executive function, including difficulties in planning, task-shifting, and self-control. Perfectionists also frequently experience rumination and worry. Such thoughts consume working memory/attentional processes, leading to encoding errors and forgetfulness, as well as slower decision-making and reduced accuracy on complex tasks. Clinical perfectionism is a strong predictor of poor cognitive flexibility, which is the ability to adapt to changing environments or task demands. Finally, a phenomenon known as "memory perfectionism" occurs when individuals have an intolerance for minor memory lapses, leading them to report severe memory problems that do not appear in objective neuropsychological tests (Desnoyers and Arpin-Cribbie 2015).

Neural substrates of perfectionism have been demonstrated in research with maladaptive perfectionism-associated structural change in the thalamus (i.e., involved in attention and memory formation among other functions) and left posterior parietal cortex (i.e., visual-spatial skills; planning and executing motor function) (Wu et al. 2017).

Functionally, maladaptive perfectionism may lead to hypervigilance for cognitive errors which, as stated previously, may result in a cycle of stress and cognitive impairment as discussed earlier in this chapter. In fact, perfectionistic personality traits may be a component of the foundation for functional cognitive disorders (FCDs) to be discussed within the next chapter. Specifically, memory perfectionism may be a driving factor in FCD creating a cycle of excessive self-monitoring of memory performance leading to anxiety over minor cognitive errors, which in turn leads to over-interpretation of normal cognitive lapses as evidence for pathological cognitive decline which then results in marked distress despite normal performance on objective cognitive testing (Mavroudis et al. 2025). This cycle of perfectionism in perception and cognitive performance may escalate over time without treatment leading to the appearance of decline over time. This cycle of negative thinking is reflected within the image below (Fig. 5.3):

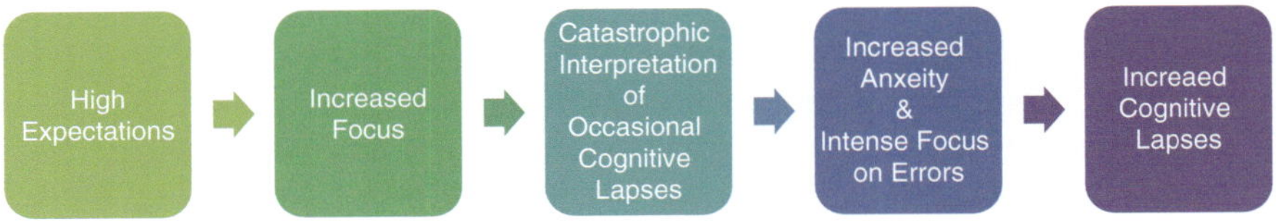

Fig. 5.3 The cycle of FCD and perfectionism

Overall, dementia is one of the most feared health conditions associated with aging, particularly in Western cultures, often ranking higher than fear for contracting cancer (Kessler et al. 2012). Such fear of forgetfulness or anxiety pertaining to the potential of dementia may be consequent to several factors including family history of dementia and/or experience with caring for patients or a loved one with dementia. Personal views and misperceptions regarding the risks and consequences of dementia, such as believing memory loss is an inevitable part of aging or that a diagnosis means life is over, contribute to the fear. This anxiety and fear can also be heightened consequent to access to biomarker testing in the context of clinical diagnostics in dementia (Graafland et al. 2025).

Such anxiety-related issues are further exacerbated by the stigma that remains in association with dementia as well as mental health stigmatization. Such concerns may include feelings about being treated differently, or losing one's identity and independence if diagnosed, are major fears associated with the social stigma surrounding both neurodegenerative and mental health disease. Perceived lack of control, specifically the feeling that there is nothing one can do to prevent or manage the illness can increase anxiety for some patients. The interconnectivity of cognitive performance with all variants of anxiety disorder further complicates diagnostic specificity and clinical management of both anxiety disorders and dementia within the context of memory clinics. Despite the complexity inherent within the relationship of anxiety and cognition, encouragingly research suggests various forms of intervention may be of significant benefit in mitigating the cognitive consequences of anxiety. This will be discussed in later chapters.

References

Abouelmagd ME, AbdelMeseh M, Elrosasy A, Eldeeb HA, Nabil Y. Adverse childhood experiences and risk of late-life dementia: a systematic review and meta-analysis. Soc Psychiatry Psychiatr Epidemiol. 2024;60(5):1087–98.

Barke A, Bode S, Dechent P, Schmidt-Samoa C, Van Heer C, Stahl J. To err is (perfectly) human: behavioural and neural correlates of error processing and perfectionism. Soc Cogn Affect Neurosci. 2017;12(10):1647–57.

Chen C, Hu Z, Jiang Z, Zhou F. Prevalence of anxiety in patients with mild cognitive impairment: a systematic review and meta-analysis. J Affect Disord. 2018;236:211–21.

Couzner L, Spence N, Fausto K, Huo Y, Vale L, Elkins S, Saltis J, Cations M. Delivering trauma-informed care in a hospital ward for older adults with dementia: an illustrative case series. Front Rehabil Sci. 2022;3:934099.

de Araujo A, Junior D, Sair HI, Peters ME, Carvalho AF, Yedavalli V, Solnes LB, Luna LP. The association between post-traumatic stress disorder (PTSD) and cognitive impairment: a systematic review of neuroimaging findings. J Psychiatr Res. 2023;164:259–69.

Desmarais P, Weidman P, Wassef A, Bruneau M, Friedland J, Bajsarowicz P, Thibodeau M, Herrmann N, Nguyen QD. The interplay between post-traumatic stress disorder and dementia: a systematic review. Am J Geriatr Psychiatry. 2020;28(1):48–60.

Desnoyers A, Arpin-Cribbie C. Examining cognitive performance: Do perfectionism and rumination matter? Personal Individ Differ. 2015;76:94–8.

Farina FR, Pavithra P, An H, Marquez M, O'Loughlin P, Regan J, Taddeo M, Bennett M, Lenaert B, Griffith JW. Validation of the Fear and Avoidance of Memory Loss scale in community-based older adults. Alzheimers Dement (Amst). 2023;15(2):e12432.

Franks KH, Bransby L, Saling MM, Pase MP. Association of stress with risk of dementia and mild cognitive impairment: a systematic review and meta-analysis. J Alzheimer's Dis. 2021;82(4):1573–90.

Graafland CH, Donker Kaat L, Richard E, van Swieten JC, Seelaar H, Bunnik EM. Ethics of disclosure of onset-predictive biomarker test results for genetic frontotemporal dementia in the research context. Alzheimers Dement (Amst). 2025;17(2):e70133.

Graham-Engeland JE, Sliwinski MJ, Almeida DM, Engeland CG. Psychological stress and cognitive brain health: policies to reduce dementia risk. Policy Insights Behav Brain Sci. 2025;12(1):94–101.

Gulpers B, Verhey F, Eussen S, Schram M, de Galan B, van Boxtel M, Stehouwer CDA, Köhler S. Anxiety and cognitive functioning in the Maastricht study: a cross-sectional population study. J Affect Disord. 2022;319:570–9.

Hill NL, Mogle J, Bell TR, Bhargava S, Wion RK, Bhang I. Predicting current and future anxiety symptoms in cognitively intact older adults with memory complaints. Int J Geriatr Psychiatry. 2019;34(12):1874–82.

Hill NL, Bratlee-Whitaker E, Jang H, Bhargava S, Sillner AY, Do J, Mogle J. Patient-provider communication about cognition and the role of memory concerns: a descriptive study. BMC Geriatr. 2023;23(1):342.

Jain N, Wang Y, Zhang Y, Jacobsen E, Andreescu C, Snitz BE, Chang CH, Ganguli M. It goes both ways: the relationship between anxiety and mild cognitive impairment. Int J Geriatr Psychiatry. 2023;38(3):e5899.

Justice NJ. The relationship between stress and Alzheimer's disease. Neurobiol Stress. 2018;8:127–33.

Karthik S, Sharma LP, Narayanaswamy JC. Investigating the role of glutamate in obsessive-compulsive disorder: current perspectives. Neuropsychiatr Dis Treat. 2020;16:1003–13.

Kessler EM, Bowen CE, Baer M, Froelich L, Wahl HW. Dementia worry: a psychological examination of an unexplored phenomenon. Eur J Ageing. 2012;9(4):275–84.

Kwak YT, Yang Y, Koo MS. Anxiety in dementia. Dement Neurocogn Disord. 2017;16(2):33–9. https://doi.org/10.12779/dnd.2017.16.2.33. Epub 2017 Jun 30.

Lachman ME, Agrigoroaei S. Low perceived control as a risk factor for episodic memory: the mediational role of anxiety and task interference. Mem Cogn. 2012;40(2):287–96.

Lynch KS, Lachman ME. The effects of lifetime trauma exposure on cognitive functioning in midlife. J Trauma Stress. 2020;33(5):773–82. https://doi.org/10.1002/jts.22522. Epub Jun 9.

Mavroudis I, Petridis F, Franekova K, Visternicu M, Rarinca V, Burlui V, Ciobica A, Novac B, Dobrin I, Hogas MM, Bovari E, Albert C, Kazis D. Functional Cognitive Disorder: diagnostic challenges, clinical features, and future directions in a misunderstood condition. Front Aging Neurosci. 2025;17:1665510.

Mohlenhoff BS, O'Donovan A, Weiner MW, Neylan TC. Dementia risk in posttraumatic stress disorder: the relevance of sleep-related abnormalities in brain structure, amyloid, and inflammation. Curr Psychiatry Rep. 2017;19(11):89.

Ostergren JE, Heeringa SG, Leon CFM, Connell CM, Roberts JS. The influence of psychosocial and cognitive factors on perceived threat of Alzheimer's disease. Am J Alzheimers Dis Other Dement. 2017;32(5):289–99.

Patterson H, Mace C, Zolnikov T, Wardlow R, Morgan S, Gordon B. A systematic review on the psychological effects of perfectionism and accompanying treatment. Psychology. 2021;12:1–24. https://doi.org/10.4236/psych.2021.121001.

Prieto S, Nolan KE, Moody JN, Hayes SM, Hayes JP. Department of Defense Alzheimer's Disease Neuroimaging Initiative. Posttraumatic stress symptom severity predicts cognitive decline beyond the effect of Alzheimer's disease biomarkers in Veterans. Transl Psychiatry. 2023;13(1):102.

Ramos ALDC, Salgado H. Obsessive-compulsive disorder as a comorbidity, risk factor or predictor of dementia. Eur Psychiatry. 2023;66(Suppl 1):S225.

Renaud C, Lacroix A. Neuroticism, perfectionism, and emotion suppression in burnout: implications for cognitive functioning. Appl Neuropsychol Adult. 2025;32(4):1138–47.

Roberts AL, Liu J, Lawn RB, Jha SC, Sumner JA, Kang JH, Rimm EB, Grodstein F, Kubzansky LD, Chibnik LB, Koenen KC. Association of posttraumatic stress disorder with accelerated cognitive decline in middle-aged women. JAMA Netw Open. 2022;5(6):e2217698.

Robinson OJ, Vytal K, Cornwell BR, Grillon C. The impact of anxiety upon cognition: perspectives from human threat of shock studies. Front Hum Neurosci. 2013;7:203.

Severs E, James T, Letrondo P, Løvland L, Marchant NL, Mukadam N. Traumatic life events and risk for dementia: a systematic review and meta-analysis. BMC Geriatr. 2023;23(1):587.

Suhas S, Rao NP. Neurocognitive deficits in obsessive-compulsive disorder: a selective review. Indian J Psychiatry. 2019;61(Suppl 1):S30–6.

Suhr JA, Kinkela JH. Perceived threat of Alzheimer disease (AD): the role of personal experience with AD. Alzheimer Dis Assoc Disord. 2007;21(3):225–31.

The Lancet Neurology. Dementia-related stigma is still pervasive. Lancet Neurol. 2024;23(11):1063.

Wu D, Wang K, Wei D, et al. Perfectionism mediated the relationship between brain structure variation and negative emotion in a nonclinical sample. Cogn Affect Behav Neurosci. 2017;17:211–23.

Perception Is Everything: Functional Cognitive Disorders

6

Prevalence of Functional Cognitive Disorder (FCD)

Functional cognitive disorder (FCD) is recognized as the cognitive counterpart to functional neurological disorders (FNDs). Functional neurological disorders, or what some may have previously called "conversion disorder," represent central nervous system symptoms thought to be consequent to the malfunction of the brain's "software" or communication pathways. Such symptoms may include cognitive change, weakness, tremors, seizures, or sensory loss, despite normal brain structure, such as reflected on neuroimaging and biomarkers. The symptoms of both FND and FCD are often triggered or worsened by stress or trauma and can be functionally debilitating to those individuals experiencing these conditions.

As a subset of functional neurological disorder (FND), functional cognitive disorder (FCD) entails symptoms specific to cognitive dysfunction. These symptoms, while very real and causing significant distress, are consequent to "functional" problems in the brain's processing rather than a progressive neurodegenerative disease like Alzheimer's disease. Estimates of FCD within clinical settings is thought to average between 20% and 30% of patients presenting to memory clinics with some studies suggesting FCD makes up a significant portion of referrals to memory clinics, especially in younger adults (i.e., under 65). Some studies note FCD may be the second most frequent diagnosis after Alzheimer's disease in memory services (Cabreira et al. 2023). Such prevalence rates are also reflected within other studies (Ball et al. 2020) suggesting FCD makes up about 24% of patients in memory clinics, with some studies finding FCD in 10 to over 50% new referrals, though prevalence varies due to varying definitions of the condition. Regardless, the prevalence patterns suggest a large portion of FCD patients presenting to memory clinics are under the age of 65 (i.e., up to 47% in one study), with variable and lesser prevalence in older adults (Cabreira et al. 2023; Borelli et al. 2022). Notably, FCD appears to afflict women more than men, often between the ages of 45 and 65 (Levine et al. 2023). Further complicating the clinical picture are the co-existing conditions for

S. Hayden, *Fear of Forgetfulness*, https://doi.org/10.1007/978-3-032-21818-6_6

patients presenting with FCD that may include chronic fatigue, fibromyalgia, depression, or anxiety (McLoughlin et al. 2023).

Generally, the variance reported in FCD frequency appears due to differing diagnostic criteria and study populations, although it is recognized as a common cause of cognitive complaints (Mavroudis et al. 2025). In essence, FCD is a relatively common diagnosis in cognitive clinics, often overlooked but increasingly recognized as a frequent reason for cognitive complaints. This has indeed become a significant component of my own practice within the Clinic for Alzheimer's Disease and Related Disorders where I would estimate one quarter to one third of the patients I am providing neuropsychological service to seem to meet criteria for FCD category.

Characteristics of Functional Cognitive Disorder (FCD)

Differentiating FCD from early dementia is crucial for optimal clinical management including avoidance of unnecessary testing which may serve to exacerbate underlying mental health issues inherent in this condition. This issue of iatrogenic (i.e., adverse consequence to medical care) exacerbation of FCD symptoms is particularly pertinent in outpatient memory clinics, where diagnostic evaluations may be required to occur over multiple years in order to provide diagnostic clarity for complex cases. This prolonged diagnostic process, although often necessary, can contribute to persistence of symptoms associated with FCD as it serves to prolong the distorted perception of an emerging dementia.

As part of diagnostic dementia evaluations in outpatient memory clinics, neuropsychological assessments exemplify the emotionally intrusive nature of cognitive testing. In the context of cognitive testing patients may often be aware and perhaps vigilant to errors made on such tasks during the testing process which then may result in increased rumination and mental health symptomatology. Thus, individuals with FCD will worry about their memory, leading to hyper-focus on even minor cognitive errors, which then overloads attention and ironically makes them less effective at performing cognitive tasks. This forms a vicious cycle of anxiety and rumination about perceived cognitive failure, similar to mental health issues and cognition described in previous chapters. The increasing cognitive errors observed in such patients may be perceived as cognitive decline and thus may be misinterpreted as an indicator of an emerging neurodegenerative disease. This process is represented in Fig. 6.1.

Within the context of possible iatrogenic processes for FCD in memory clinics via diagnostic evaluations, the key characteristics and symptoms of FCD are in themselves complex and include the following considerations:

- *Subjective Complaints Versus Objective Function*: A key feature of FCD is the discrepancy between the patient's subjective perception of severe cognitive decline versus their performance on objective neuropsychological tests, which is often within normal limits or shows inconsistencies in cognitive performance. This is well demonstrated within the discrepancy noted between patient

Fig. 6.1 Functional cognitive disorder cycle

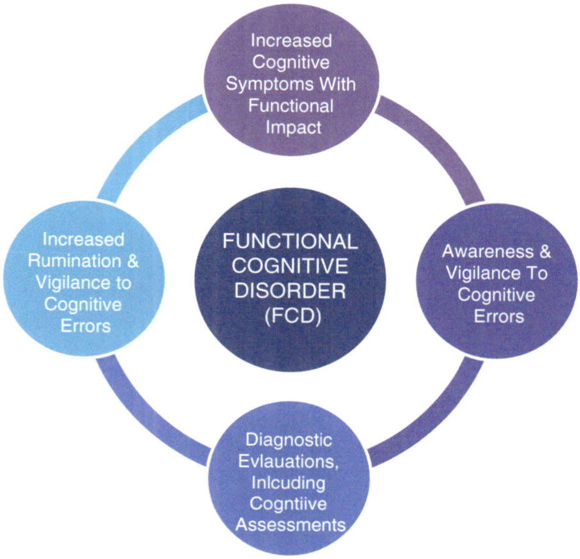

interview reflecting perceived cognitive performance and their actual performance on detailed neuropsychological testing.

- *Heightened Self-Monitoring/Hypervigilance*: Individuals with FCD often engage in "memory perfectionism" which contributes to a self-perpetuating cycle where individuals excessively monitor their own memory/cognitive performances (i.e., hypervigilance) leading to further anxious self-monitoring of their cognitive performance. This constant focus on potential memory failures overloads their attention and working memory, paradoxically making the cognitive symptoms worse (i.e., leading to increased forgetfulness).
- *Illness Perception*: A patient's perception or misperception of believing they have a progressive, degenerative disease like dementia will cause significant anxiety and distress, which in turn worsens their cognitive symptoms.
- *Vicious Cycle*: The perception of severe cognitive impairment may lead to a lowered or lack of confidence in their own memory performance, causing individuals with FCD to rely heavily on aids or other people, which then leads to a de-skilling of their own memory and a worsening of symptoms, thus creating a "vicious cycle."

The Role of "Perception" in FCD

The phrase "perception is everything" directly relates to FCD given such patient's strong belief and heightened awareness (i.e., hypervigilance) for their cognitive lapses. This issue entails impairments in functions that may create self-fulfilling prophecy including issues with the following:

- *Metacognition*—This is an individual's awareness or knowledge of their own cognitive processes (i.e., "thinking about thinking") as well as the regulation/control of these cognitive processes. This allows an individual to plan, monitor, and evaluate their own actions or learning.
- *Negative Interpretation Bias*—This refers to a cognitive bias involving interpretation of ambiguous situations or information in a negative or threatening manner, rather than in a neutral or positive way, regardless of the actual threat level.

Commonly, in my own clinical practice, I may notice such negative interpretation biases and/or dysfunction of metacognition within the initial clinical interview when asking patients about their cognitive concerns. Although such clues to FCD may be subtle, a skilled clinician can attend to the indicators of such perfectionistic tendencies. This may also be witnessed during the process of cognitive testing, through commentary or behavior expressed by FCD patients (i.e., excessively negative evaluation of performance on tasks, despite adequate performance levels) while undergoing neuropsychological evaluation.

Further, regarding the concept of perception for FCD patients, this "disconnect" between patients' perception of cognitive symptoms relative to their measurable cognitive performance can result in considerable challenges, and in some cases conflict for clinicians caring for FCD patients. To optimize care, clinicians must recognize FCD as a significant potential differential diagnosis within memory clinics, as supported by the aforementioned evidence of FCD prevalence within these clinical settings. When navigating the misperceptions of FCD patients, clinicians must approach the disconnect between perception and measured performance with empathy and support, rather than dismissiveness. Acknowledging and providing education regarding a patient's cognitive concerns distinct from a diagnosis of dementia is essential for FCD patients as with other mental health issues discussed previously. To be able to effectively provide such education regarding the nature of a patient's cognition, a comprehensive evaluation of both cognitive and psychological factors is warranted, which is often within the scope of neuropsychology.

Of note, although the etiology of FCD is unclear, Teodoro et al. (2018) have proposed a unifying hypothesis for the mechanisms underlying the cognitive issues associated with FCD. They have proposed factors such as *excessive interoceptive monitoring* (i.e., maladaptive heightened attention to cognitive processes) in FCD leads to a decrease in externally directed attention. This then increases susceptibility to distraction and slows information processing, interfering with other aspects of cognitive function, particularly executive functions (i.e., multitasking) and encoding for memory. For FCD patients, routine cognitive processes are consequently experienced as unduly effortful. These experiences might then be overinterpreted as evidence of dementia due to memory perfectionism and heightened self-monitoring of cognitive performance.

Clinical Implications for FCD

When clinicians, including neuropsychologists, are conducting assessments with FCD patients in memory clinic settings, there is a need to consider a variety of issues when examining the behavioral presentation and cognitive profiles of these patients including the following:

- *Subjective Complaints Versus Objective Performance*: As mentioned previously, FCD patients presenting to memory clinics typically report significant forgetfulness along with possible issues in attention and language as well as other aspects of cognition. However, this perception of cognitive performance is markedly inconsistent with performances on formal cognitive tests which are often within expected range for age.
- *Internal Inconsistency*: Performance on cognitive tasks with FCD patients may be internally inconsistent, reflecting variability in performance outcomes evident on various tests within the *same* cognitive domain. For instance, a person might struggle with a verbal memory task at one moment but subsequently perform relatively well on another verbal memory task at another time. Behavioral observations are critical in order to identify specific behavioral reactions observed during these periods of variability.

  Further, there may be evidence of significant variance in degree of difficulty performance. In such instances, FCD patients may perform poorly on simpler versions or aspects of a test versus more complex aspects of specific tasks (i.e., FCD patients may demonstrated better delayed recall than immediate recall on memory tasks which would be inconsistent with that expected for dementia).
- *Preserved Insight and Detail*: Unlike many dementia patients, individuals with FCD usually have reasonable insight into their difficulties and awareness of perceived cognitive difficulties. FCD patients may often provide very detailed, coherent descriptions of their memory failures and in some cases may come to Clinic with detailed written notes on such cognitive errors. This is in contrast to a subset of patients with likely dementia who will expressly state they do not have concern for their cognitive function and may even question why they are being examined.
- *Attending Alone*: Patients often attend medical/neuropsychological appointments alone, with no apparent issues in traveling to such appointments independently, whereas those with neurodegenerative conditions more frequently bring a family member or caregiver to provide collateral information. In other words, FCD patients may describe marked cognitive deficits that would be expected to significantly impede independent function, including driving or taking the bus on their own, that are inconsistent with their independent attendance to the clinic appointment.
- *Associated Features*: FCD is commonly associated with comorbidities such as anxiety, depression, chronic pain, and sleep disturbances, which can all exacerbate cognitive symptoms. Such symptoms may be measurable on psychological testing and/or may present themselves within clinical interviews.
- *Positive Diagnosis*: Finally, clinicians working with FCD patients should aim to use the above specific patterns of symptoms and illness perception as "positive diagnostic clues" to identify FCD, rather than just a diagnosis of exclusion as is typical when making diagnosis of dementia.

Although the clinical presentation of FCD is certainly complex within the context of differential diagnosis from neurodegenerative conditions, perhaps the representation below on the distinction of diagnostic features between FCD and dementia provided in Table 6.1 may give some clarity (Table 6.1).

Within clinical practice, I have had many patients meeting all or most of the aforementioned criteria when presenting to our memory clinic. Careful and comprehensive cognitive assessment accompanied by in-depth psychological testing has informed my own evaluation of FCD patients. This form of assessment, as detailed within Chap. 2, can provide information essential to the differential diagnosis of FCD versus dementia. Such detailed measurement may be useful to the neuropsychologist as well as subsequent clinicians in communications with FCD patients about their perceived cognitive performances. The use of concrete data secured through neuropsychological testing can be an effective tool to mitigate some of the misperceptions and related conflict that may arise with FCD patients who resist any findings that their measured cognitive performance is intact. In my experience with FCD patients, there is a subset of these individuals who will become exceedingly defensive and/or angry toward a clinician reporting their intact cognitive performances to them. It may indeed seem counterintuitive to a clinician that such patients would rather be informed they have FCD than dementia, but this likely reflects societal and self-stigmatization discussed previously.

Commonly in my own practice, I begin the follow-up neuropsychological feedback session regarding a patient's test result with a query on how the individual perceived their own cognitive performance on testing several weeks prior. For FCD patients, they will typically recall their test performance without issue and will be prone to overestimate errors within the cognitive testing as well as underestimate mental health factors based upon their response to psychological testing. It is from this starting point that discussion of the disconnect between perception of cognition

Table 6.1 Distinguishing diagnoses: FCD versus dementia

Feature	Functional cognitive disorder (FCD)	Dementia (i.e., Alzheimer's disease)
Primary cause	Functional dysfunction in how brain networks process attention and monitor memory	Neurodegenerative disease (e.g., plaques & tangles) causing brain cell loss
Progression	Generally nonprogressive; symptoms may fluctuate over time but do not consistently worsen	Progressive symptoms worsen over time as the underlying disease spreads
Testing	Often shows internal inconsistency (i.e., performing well one moment and poorly the next; or performing better on simpler vs. more challenging tasks)	Shows consistent deficits across formal tests, which may be reflected in daily life
Awareness	High self-awareness regarding cognitive issues. FCD patients are often very worried about their memory, and in some cases more so than their loved ones	Often marked by a lack of insight (i.e., anosognosia); family members may be more worried than the patient is about their cognition
Interaction	Patients often attend appointments alone and may bring detailed written notes regarding cognitive function	Patients are usually accompanied by family on whom they may rely on to answer questions for them

and the measured performance may begin when communicating results. Such a conversation may go like this:

Neuropsychologist: *So, do you recall the day of testing that we completed a few weeks ago?*

Patient: *Yes, I do remember....it was a very long day.*

Neuropsychologist: *Indeed, it was a long day, and we appreciated the work you put in that day. Before I discuss the results of that day of testing, I am curious how you felt it went. Can you tell me how you thought your performance was on any or all of the testing that was conducted?*

Patient: *Oh, I thought I was horrible on those memory tests...I could not memory anything on those tests at all, although it is actually not surprising to me that I was so bad at those tests.*

Neuropsychologist: *Ok well, let's talk about that then. I would start with the good news that actually, for your age and education, your performances all fall into average range. Now, this doesn't mean that you have not been struggling on a day to day basis but it does give us clues to what might be happening. Also, although I know it may be surprising but the difference between how you felt you did on testing and how you actually performed on these tests can give us another clue to what is going on. Yet another clue lies in your responses on the testing for mood and personality which revealed high anxiety and somatization symptoms. All these clues are pointing to the possibility of a functional cognitive disorder, which can certainly cause significant and for some debilitating cognitive symptoms on a variable basis. This variability was seen in your testing as well (give examples).*

At this point in the feedback, I would encourage questions from the patient and their families (if present) regarding what I had just discussed. I then would provide full discussion of the meaning of functional cognitive disorder as well as in-depth discussion of the cognitive performance and psychological test findings evident on the neuropsychological assessment. Although there is a fine balance to navigate with individuals prone to rumination as in FCD, providing clear and open communications regarding the relationship between their functional cognitive performances and the underlying mental health factors is critical to advancing care for such patients. Such clinical communications often take time to ensure a comprehensive review of symptoms and their meaning in the context of FCD.

Role of Complementary Investigations Neuroimaging or other medical tests are not typically needed for FCD per se, but such investigations may be useful if there is a suspicion of a comorbid or contributing medical condition. Although, one certainly needs to avoid unnecessary testing in all aspects of medicine, some testing may provide further clarification to both clinician and patient regarding etiologies of cognitive concern. As discussed in prior chapters, this issue is particularly complex in the context of biomarker testing. Biomarkers (i.e., blood tests for p-tau, NfL, Amyloid) and advanced imaging help distinguish FCD from neurodegenerative disorders by elucidating underlying pathology (or lack thereof) in dementia in some cases, guiding diagnosis and treatment. Biomarker tests, especially new blood tests for proteins like ptau217, are improving to detect early dementia, but a combination of clinical assessment, cognitive tests, imaging, and biomarkers may prove to be essential for accurate diagnosis, differentiating from FCD from neurodegenerative conditions. In FCD, biomarkers may be normal or show nonspecific changes, while dementia biomarkers would reveal specific Alzheimer's or other neurodegenerative pathology. The inconsistent symptoms and relative preservation of measured cognitive function in FCD patients contrast with dementia's progressive decline, with biomarkers helping to confirm the presence or absence of neurodegeneration.

Of note and to further complicate diagnostic process however, FCD can also co-occur with Alzheimer's disease which has been referred to as FCD-AD involving cognitive symptoms that may overlap and possibly precede Alzheimer's disease presentation (Cabreira et al. 2025). These researchers cite the comorbidity of functional neurological disorders and neurologic disease, with the suggestion that FCD may be a prodromal feature of AD although this relationship is most certainly not definitive.

Management/Treatment of Functional Cognitive Disorders

The management of FCD is complex and treatment must be individualized to patient needs given the complexity of this condition. Ideally, treatment for FCD is multidisciplinary, focusing on retraining the brain through specialized occupational therapy, psychotherapy (i.e., such as Cognitive Behavioral Therapy—CBT), and stress reduction. Such interventions are often combined with the clinical management of co-occurring conditions like anxiety or depression with medication, aiming to reverse symptoms and improve function (Van Patten and Keatley 2025). Patient education and engagement is key to positive clinical outcomes. Below is a list of the treatment modalities to be considered when working with FCD patients.

- *Psychoeducation and Clinical Communications*: Simply providing the correct diagnosis and explaining the condition (i.e., psychoeducation) is often an effective first step in clinical management of FCD patients. Such communications serve to increase the FCD patient's understanding of the reasons for their distressing functional cognitive symptoms and provide opportunity to discuss potentially reversibility of such symptoms. Ideally, this would be within the context of early identification and diagnosis of FCD, in order to prevent unnecessary, costly, and potentially stigmatizing investigations for dementia.

  With respect to patient-clinician communications for FCD diagnoses, it is important to explain that diagnostics for FCD is based on patterns present within the objective testing along with their own clinical history, communications, and behavior, rather than just a single test. Many patients may find this approach to assessment perplexing and may question its validity. The responsibility, therefore, is upon the clinician/neuropsychologist to take the time necessary to provide appropriate psychoeducation on all aspects of the FCD patient's presentation to circumvent further distortions of perception.

- *Psychotherapy*: Treatments targeting FCD must often involve addressing the underlying factors, such as anxiety, depression, trauma, and poor sleep. The use of psychological treatments like Cognitive Behavioral Therapy (CBT) to alter maladaptive illness perceptions and self-monitoring behaviors may be effective. Once again, preceding any proposal for such treatment, a full discussion regarding the role of mental health factors in cognitive performance as well as in the presentation of FCD is required. Further of note, when I am recommending CBT treatments in such cases, I specifically indicate it should be conducted by a psychologist with experience in functional cognitive disorders or functional neurological disorders. As FCD is complex, any treating mental health provider must have some working knowledge of the clinical presentations of this condition.
- *Metacognitive Retraining*: A form of cognitive-behavioral therapy approach is metacognitive retraining which helps FCD patients increase their awareness of misperceptions and supports change in these maladaptive thinking patterns (i.e., cognitive biases), reducing symptoms of FCD. This form of therapy assists FCD patients in better directing their attention in order to improve cognitive efficiency and reduce cognitive errors on a functional basis.
- *Acceptance and Commitment Therapy (ACT)*: ACT is a mindfulness-based behavioral therapy that can support FCD patients in the development of psychological flexibility by accepting difficult thoughts and feelings instead of fighting them, while committing to actions that align with their core values to build a more meaningful life, rather than focusing on symptom reduction.
- *Lifestyle Modifications*: Improving sleep hygiene, engaging in regular physical activity, and adopting a healthy diet (e.g., Mediterranean, MIND) can be encouraged in FCD patients to provide a proactive engagement in health habits.
- *Mindfulness and Relaxation*: Techniques including deep breathing, progressive muscle relaxation, yoga, and biofeedback help manage stress and physical tension associated with FCD.
- *Managing Underlying Conditions*: Treating comorbid depression, anxiety, chronic pain, or fatigue is crucial for reducing cognitive "fog" for patients with FCD.
- *Occupational Therapy*: Occupational therapists can provide guidance on functional strategies to manage cognition in daily life, including the use of memory aids (calendars, lists) and pacing activities to avoid negative performance cycles of FCD.
- *Medication Review*: A healthcare provider (i.e., neurologist or psychiatrist) may need to review, adjust, or stop medications that might be contributing to cognitive impairment for patients with FCD and/or may need to consider medications to manage underlying mental health issues.
- *Self-Help Tools*: Tools like the *Mementum* app which is a self-help app to "empower individuals with Functional Cognitive Disorder (FCD)." This app includes learning modules as well as options for weekly check-ins.

- *Patient Support*: Management of FCD patients also involves individualized, compassionate care, including support for the patient and their caregiver, and encouraging activities that support brain health, to be discussed in later chapters of this book.

Overall, the diagnosis and management of FCD is complex and multifactorial with need for further research to advance our understanding of underlying etiologies in order to better predict and cognitive course in these patients and to optimize treatment outcomes. Given that the prevalence of FCD diagnoses may represent 20 to 30 percent of patients presenting to memory clinics for investigation of dementia, advancing the understanding of this condition within the context of neurodegenerative conditions is critical. This is also a pressing need for further understanding in this area given the paucity of research regarding FCD and dementia co-morbidities. Clearly, the approach to diagnostic assessment and management of FCD differs markedly from that of traditional dementia evaluation. Consequently, clinical resource appropriate to FCD detection and management must be included within the memory and dementia clinics. Without this expansion of scientific understanding and associated clinical resource regarding FCD diagnoses within dementia clinics, misdiagnosis and iatrogenic sequelae is heightened.

References

Ball HA, McWhirter L, Ballard C, Bhome R, Blackburn DJ, Edwards MJ, et al. Functional cognitive disorder: dementia's blind spot. Brain. 2020;143(10):2895–903.

Borelli WV, de Senna PN, Brum WS, Schumacher-Schuh AF, Zimmer ER, Fagundes Chaves ML, Castilhos RM. Functional cognitive disorder presents high frequency and distinct clinical profile in patients with low education. Front Aging Neurosci. 2022;14:789190.

Cabreira V, Frostholm L, McWhirter L, Stone J, Carson A. Clinical signs in functional cognitive disorders: a systematic review and diagnostic meta-analysis. J Psychosom Res. 2023;173:111447.

Cabreira V, Zeidler M, McWhirter L, Suvankar P, Stone J, Carson A. Functional cognitive disorder in Alzheimer's disease. J Neurol Sci. 2025;476:123618.

Levine DA, Gross AL, Briceño EM, Tilton N, Giordani BJ, Sussman JB, Hayward RA, Burke JF, Hingtgen S, Elkind MSV, Manly JJ, Gottesman RF, Gaskin DJ, Sidney S, Sacco RL, Tom SE, Wright CB, Yaffe K, Galecki AT. Sex differences in cognitive decline among US adults. JAMA Netw Open. 2023;4(2):e210169. https://doi.org/10.1001/jamanetworkopen.2021.0169. Erratum in: JAMA Netw Open. Mar 1;6(3):e234786.

Mavroudis I, Petridis F, Franekova K, Visternicu M, Rarinca V, Burlui V, Ciobica A, Novac B, Dobrin I, Hogas MM, Bovari E, Albert C, Kazis D. Functional Cognitive Disorder: diagnostic challenges, clinical features, and future directions in a misunderstood condition. Front Aging Neurosci. 2025;17:1665510.

McLoughlin C, Hoeritzauer I, Cabreira V, Aybek S, Adams C, Alty J, Ball HA, Baker J, Bullock K, Burness C, Dworetzky BA, Finkelstein S, Garcin B, Gelauff J, Goldstein LH, Jordbru A, Huys AM, Laffan A, Lidstone SC, Linden SC, Ludwig L, Maggio J, Morgante F, Mallam E, Nicholson C, O'Neal M, O'Sullivan S, Pareés I, Petrochilos P, Pick S, Phillips W, Roelofs K, Newby R, Stanton B, Gray C, Joyce EM, Tijssen MA, Chalder T, McCormick M, Gardiner P, Bègue I, Tuttle MC, Williams I, McRae S, Voon V, McWhirter L. Functional neurological disorder is a feminist issue. J Neurol Neurosurg Psychiatry. 2023;94(10):855–62.

Teodoro T, Edwards MJ, Isaacs JD. A unifying theory for cognitive abnormalities in functional neurological disorders, fibromyalgia and chronic fatigue syndrome: systematic review. J Neurol Neurosurg Psychiatry. 2018;89(12):1308–19.

Van Patten R, Keatley E. Cognitive rehabilitation for functional neurological disorder. CNS Spectr. 2025;31(1):e1.

Context Matters: Psychosocial Factors and Cognition

7

Psychosocial Factors in Cognition

Psychosocial factors within the context of dementia diagnostics refer to an individual's psychological characteristics and their consequent interactions within a social environment that may significantly impact risk for cognitive decline. Such factors have been examined by scientists, including the researchers Rehan and Phillips (2025) who explored associations between psychosocial function and cognitive performance in multiple domains for individuals with mild cognitive impairment (MCI). The findings of this study suggested social participation (i.e., degree of engagement in social activities) was uniquely associated with cognitive performance on neuropsychological tests in the areas of verbal fluency and processing speed. However, there were no associations between psychosocial factors and cognitive performance on measures of memory, executive function, and working memory/attention. Interestingly, this pattern of cognitive impact by psychosocial factors, with greater compromise in attention and processing speed, appears to mimic areas of cognition typically affected by mental health disorders. Such convergent effects for psychosocial and mental health factors upon cognition only further highlights the necessity for in-depth understanding of both of these mitigating factors within the clinical care of patients being investigated within memory clinics.

Although further exploration of the role of psychosocial factors is certainly necessary, it is important to acknowledge that a patient's individual context matters a great deal when interpreting cognitive performances, as discussed earlier in this book. It is helpful to consider such psychosocial factors in terms of risk of cognitive decline, specifically both positive and negative risk factors, as illustrated in Table 7.1:

Interplay of Factors

Psychosocial and contextual factors are deeply interconnected with cognitive functional performance (Li et al., 2025). The influence of these factors on cognitive

S. Hayden, *Fear of Forgetfulness*, https://doi.org/10.1007/978-3-032-21818-6_7

Table 7.1 Psychosocial factors in prediction of cognitive health

Positive psychosocial factors (*protective*)	Negative psychosocial factors (*risk*)
Social Engagement: Frequent social contact including community involvement and strong social networks are linked to lower dementia risk and better cognitive performance	*Chronic Stress & High Demands*: Unmanaged stress such as poor work-life balance is associated with lower cognitive performance and cognitive decline, specifically in memory and executive function
Mental Stimulation: Engaging in mentally stimulating activities builds cognitive reserve and supports resilience to cognitive decline	*Depression & Anxiety*: These are significant factors in cognitive presentation as they may exacerbate or worsen cognitive symptoms
Psychological Well-Being: Optimism, self-efficacy, purpose in life, and good relationships all are likely to be supportive of cognitive health	*Loneliness & Isolation*: Subjective feelings of loneliness and actual social isolation increase risk for cognitive decline
Perceived Control/Self-Efficacy: Feeling in control of one's life is associated with better working memory and cognitive resilience	*Low Socioeconomic Status*: Socioeconomic adversity leads to stress which can diminish cognitive function
Sense of Purpose: Optimism, purpose in life, self-efficacy, and good relationships all are likely supportive of cognitive health	*Stressful Events*: Adverse childhood psychosocial factors, including stress, can affect midlife learning and memory.
Education: Higher education is associated with cognitive reserve	*Lower Educational Attainment*: This is associated with higher likelihood of cognitive decline

decline is often bidirectional, as has been the case for mental health factors discussed in previous chapters. Thus, poor psychosocial function can lead to cognitive decline, but cognitive impairment can also reduce social engagement, creating a cycle as illustrated below (Fig. 7.1).

For example, low socioeconomic status can lead to increased financial strain, a psychosocial stressor, which in turn may lead to chronic stress and a higher risk of cognitive decline. This may be further exacerbated should such psychosocial factors be sustained long term with increased risk for the development of mental health disorders. As discussed in prior chapters, mental health disorders then carry additional risk for cognitive dysfunction and in some cases risk for decline. Positive factors and resilience may mitigate risks for cognitive decline. For instance, strong control beliefs (i.e., an individual's perceptions of the presence or absence of factors that can facilitate or impede the performance of a behavior) can offset the negative cognitive impact of low education as a risk for cognitive deterioration. Recognizing these complex interactions is crucial for developing effective, targeted interventions to preserve cognitive health across the lifespan. However, this understanding seems particularly pertinent within the context of diagnostics and clinical care for dementia. Without fulsome exploration of psychosocial factors within the clinical assessments for neurodegenerative conditions, there is a risk for misdiagnosis as well as the potential for overlooking mitigation of such factors.

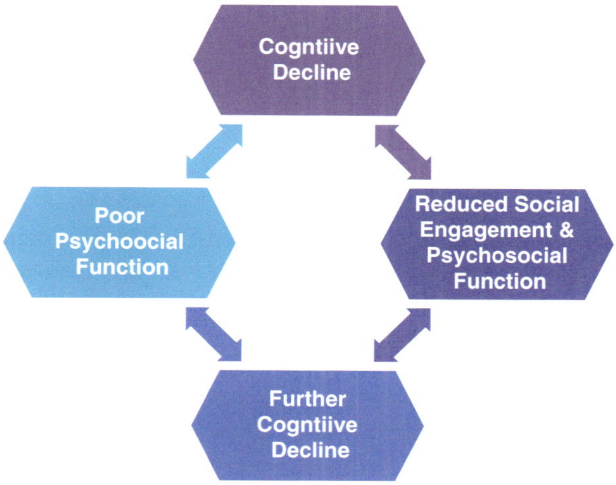

Fig. 7.1 Interaction of psychosocial factors and cognitive decline

Psychosocial Factors in Brain Structure and Function

Psychosocial factors, such as those discussed above, can directly influence brain structure and function, which further impacts various cognitive domains like attention and processing speed, which in turn may result in functional impact on co-dependent cognitive functions such as memory and aspects of executive skill. Psychosocial factors ranging from early-life adversity (i.e., trauma) to chronic stress to social engagement to purpose in life all exert significant influence on brain function by altering the brain's molecular biology, structural integrity, and cognitive capacity. The mechanisms of brain structure and function impacted by psychosocial factors may be best categorized into the following areas.

Molecular and Biological Pathways

- *Mitochondrial Function*: Research indicates a strong relationship between psychosocial function and mitochondria which are the "powerhouse" cells designed to generate energy (i.e., ATP) used for body functions. Mitochondria are considered sensors to stress that translate our social experiences to biological changes. Positive psychosocial experiences are linked to a higher abundance of mitochondrial in the brain. Conversely, negative psychosocial factors are associated with lower mitochondrial energy capacity, which may drive cognitive decline, according to some research (Picard and McEwen 2018; Trumpff et al. 2024).
- *Neuroendocrine (HPA) Axis*: Negative psychosocial factors, such as psychosocial stress, trigger the sustained release of cortisol via the hypothalamic-pituitary-adrenal (HPA) axis. Over time, this can lead to "allostatic load" which refers to the physiological wear-and-tear that disrupts the brain's ability to regulate stress and emotions (McEwen and Gianaros 2010).

- *Glutamate Dysregulation*: Negative psychosocial factors such as prolonged stress can cause unregulated glutamate overflow, which has neurotoxic effects and may represent a key driver in brain aging and dementia (Vaidya et al. 2024).

Structural and Functional Changes

Psychosocial factors impact the brain's *neuroplasticity* which essentially entails the brain's ability to reorganize its structure, functions, and neural connections throughout life in response to learning, experiences, environmental changes, or injury, allowing it to adapt, form new skills, create memories, and recover from damage. For example, some research suggests (Zannas 2018) gene-stress interactions may influence the aging brain and contribute to vulnerability or resilience to neuropsychiatric syndromes including anxiety, depression, and FCD. Such research suggests there may be a direct impact of psychosocial upon brain structure, which in turn will have significant impact upon cognitive and behavioral aging (Wilson and Bennett 2017). The role of psychosocial factors in neuroplasticity is reflected within measurable changes in specific regions:

Limbic System (Amygdala and Hippocampus): Psychosocial factors appear to significantly influence the limbic system, a network of brain structures that regulates emotions, memory, and the stress response. In essence, the limbic system acts as a central hub, where the brain translates psychosocial experiences into physiological and emotional responses. More specifically, the following psychosocial factors have been associated with measurable change in the amygdala and hippocampus regions:
  - *Childhood Adversity*: Childhood trauma, maltreatment, or chronic stress often leads to a hyperactive amygdala (i.e., involved in the fear response or what is referred to as "fight-flight") and reduced volume in the hippocampus (i.e., involved in memory and learning).
  - *Social Isolation*: Loneliness is associated with smaller volumes in the hippocampus and amygdala, potentially accelerating cognitive aging and raising risk for cognitive decline (Zhang et al. 2018).

Prefrontal Cortex (PFC): Psychosocial factors significantly influence the structure and function of the prefrontal cortex (PFC) during aging, affecting how individuals cope with stress, regulate emotions, and maintain cognitive function over time (Zahodne 2021). While the PFC naturally undergoes age-related decline, positive psychosocial factors can build cognitive resilience, while negative factors can accelerate cognitive deterioration. Negative psychosocial factors such as chronic stress and low socioeconomic status (SES) can shrink the PFC, which is the region responsible for executive function, decision-making, and emotional regulation.

White Matter Integrity: Negative psychosocial factors, such as high stress levels, across the lifespan are linked to reduced microstructural integrity in white matter tracts, which impairs communication between brain regions and in turn may reduce cognitive efficiency.

Impact on Cognitive Function

As mentioned, emerging evidence suggests the direct impact of psychosocial factors on brain regions and functions that will undoubtedly result in cognitive dysfunction as we age and may impact risk for neurodegenerative factors. Negative psychosocial factors such as social isolation and depression have been found to be robust predictors of a more rapid rate of cognitive decline and a higher risk of developing dementia (Shen et al. 2022). In contrast, high cognitive reserve may be built through positive psychosocial stressors such as higher educational attainment, social engagement, and a sense of purpose, all of which can buffer the brain against age-related pathologies. For example, older adults with large friend networks show a weaker link between brain atrophy and cognitive impairment. Further, enhancing positive psychosocial factors are presenting promising potential for improving cognitive function with well-being (i.e., living one's life in keeping with one's potential) associated with better cognitive health and lower incidence of cognitive deficit and neurodegenerative disease (Masurkar et al. 2024).

Overall, fostering positive psychosocial health through social connections, mental activities, and emotional well-being is crucial for building cognitive reserve and promoting healthy cognitive aging. In the many patients I have served in my career, the issue of psychosocial context has repeatedly proven to be an important matter in the analysis of neuropsychological cognitive performance.

Psychosocial Factors in Functional Cognitive Disorders

Psychosocial factors are central to functional cognitive disorder (FCD), and thus warrant further review within this chapter. As with various mental health disorders discussed previously, psychosocial factors play a role in the expression and maintenance of FCD symptoms (Cabreira et al., 2023). The psychosocial factors impacting the presentation of FCD are typically acting via three possible mechanisms including the following:

- *Predisposing Factors*—which involve factors that increase vulnerability for FCD
- *Precipitating Factors*—entail factors that trigger symptoms of FCD
- *Perpetuating Factors*—refer to factors that sustain FCD symptoms

All of these factors contribute to the significant subjective cognitive complaints reported by FCD patients, despite normal objective test performance. This serves to highlight the links between mental health, psychosocial factors, stress, and cognitive experience (Patron et al., 2022; Pennington et al. 2019). Below is a summary of these subtypes of psychosocial factors that play a role in FCD patient presentation.

Predisposing (Vulnerability) Factors These factors may increase an individual's vulnerability to functional cognitive disorders (FCDs). Although such factors are not predictive of FCD (i.e., not guaranteed), it raises the risk or potential for development of this disorder. The predisposing factors discussed below largely reflect psychological characterstics, which in turn, would have impact upon social and functional factors over time.

- *Mental Health*: There is a high prevalence of co-existing mental health issues that are associated with increased vulnerability to FCD symptoms (du Pont et al., 2019). Such mental health issues include anxiety, depression, perfectionism, and obsessive-compulsive traits, all of which have been discussed in greater detail within previous chapters. Nevertheless, these mental health comorbidities have been estimated to co-occur at a rate of 51%–95% in functional cognitive disorders (FCDs) which may be a contributing factor to the under-diagnosis of FCD within memory clinics focussed upon dementia diagnostics (Mavroudis et al. 2025a, b). Thus, awareness of and treatment for such comorbidities is essential in the diagnoses and effective management of FCD.
- *Neurodevelopmental Traits*: There is an overrepresentation of autism spectrum (ASD) characteristics and attention deficit/hyperactivity disorder (ADHD) symptoms in FCD populations (Staunton et al. 2024). The neurodevelopmental traits associated with ASD and ADHD are considered predisposing factors to and are prevalent in FCD populations, potentially reflecting sensory processing anomalies and predictive of coding impairments within their presentation. Once again, such co-morbidities serve to inadvertently cloud the diagnostic process for FCD within the context of memory clinics.
- *Personality*: Cognitive rigidity or tendencies to misinterpret normal lapses as serious problems (i.e., perfectionism, tendency to ruminate) are associated with increased vulnerability to FCD and in fact represent key characteristics of individuals presenting with this disorder (Teodoro et al. 2018). Personality characterstics, specifically those meeting criteria for a personality disorder, may be more challenging to effectively treat and thus may represent a greater challenge to modfication as a risk factor to FCD.

Precipitating (Triggers) Factors These factors are not related to causality in FCD patients but rather refer to issues that act as a catalyst or trigger for expression of FCD symptoms. Such precipitating factors are often important to identify within clinical context, so that FCD patients may be educated on their specific triggers and the consequent impact upon FCD symptoms. This represents an important step in the management/treatment of FCD symptoms.

- *Stressful Events*: Trauma, significant life stressors, or recent psychological disruptions may act as triggers to FCD onset (Cabreira et al. 2025). As mentioned previously, discussing stressors and trauma, with patients presenting cognitive concerns for dementia investigation is critical for optimal clinical care. This is particularly critical given that FCD is widely under-diagnosed within such clinical settings.
- *Physical Illness*: Viral infections, chronic fatigue, or sleep disturbances may also trigger FCD symptoms (Tackey et al. 2024). Such triggers are particularly

important given the ongoing pursuit of understanding long COVID symptoms. Integrating queries regarding past experiences with infections, including COVID-19, and other physical illness can not only provide information into such triggers but also may give further insight into perceptions patients may hold in response to illness or disease. Although full understanding of the long-term effects of persistent COVID-19 symptoms requires further research, clinical documentation of such correlations may support future understanding of such precipitating factors.
- *Burnout*: Occupational burnout can increase vulnerability and may be a trigger for FCD (Chalak et al. 2022). Chronic workplace stress can lead to emotional exhaustion and 'burnout' which can subsequently result in symptoms similar of FCD including issues with memory loss, difficulty focusing, and poor executive function. Identification of factors such as burn out is critical to ensure such triggers can be mitigated. Further, management of burn out and it's sequelae may lead to greater diagnostic specificity for conditions involving cognitive dysfunction such as FCD and/or dementia.
- *Iatrogenic Factors*: Ambiguous diagnostic labeling provided by clinicians or perceived invalidation by healthcare providers can act as stressors that precipitate or worsen symptoms of FCD. Unfortunately, given the under-diagnosis and misunderstanding of FCD in memory clinics, it is not uncommon for such vague or dismissive communications to occur. As mentioned in past discussion of FCD, comprehensive neuropsychological evaluation can be a very useful tool in mitigating such iatrogenic factors related to communciations regarding FCD symptoms.

Perpetuating (Maintenance) Factors These factors do not cause or trigger FCD symptoms but serve to maintain the symptoms or allow the symptoms to persist over time. Given the role of symptom maintenance for FCD, it is important for FCD patients to learn skills, often through psychotherapy, to mitigate these perpetuating factors. Furthermore, clinicians must become better aware of the role of their own clinical communications in the maintenance of FCD symptoms.

- *Dysfunctional Beliefs*: Fear of neurodegeneration or dementia despite normal test results (e.g., "brain fog"), *without* clarification and psychoeducation about the cause of symptoms can maintain symptoms of FCD. This, once again, is why effective communication regarding symptoms and how mental health factors impact functional cognitive performance is critical to circumventing the symptoms of FCD on an ongoing basis.
- *Social Factors*: Loneliness, as well as low social support, and reduced social participation, which are all also risk factors for general cognitive decline, can also worsen and maintain FCD symptoms over time (Guarnera et al. 2023). Thus, social supports and social programs represent critical interventions for FCD, as they do for dementia populations.

- *Cogniphobia*: This represents a behavioral pattern of avoidance for cognitively demanding tasks due to an individual's fear of failure. This avoidance prevents exposure to corrective experiences and consequently serves to sustain FCD. Psychological treatment including exposure therapy (i.e., a component of Cognitive Behavioral Therapy-CBT which entails exposing patients to feared tasks/circumstances in a controlled environment) is often an effect means of managing this perpetuating factor of FCD.
- *Healthcare Experiences*: Excessive investigations and/or poor communication with clinicians may lead to reinforcement of maladaptive beliefs serving to maintain symptoms of FCD. Unfortunately, given limited understanding and experience with FCD in memory clinics, many clinicians may recommend repeated medical investigation that only serve to reinforce a patient's fear of an underlying disease, such as dementia. This issue is further exacerbated by poor or dismissive communications by clinicians which only lead to an increase in anxiety and the belief that symptoms are "unexplainable" and ominous which is a hallmark mistaken belief of FCD. Clearly, improvement in the education and skill development of clinicians for recognition and communications surrounding FCD is required, particularly for those working within memory clinics.

Overall, the predisposing, precipitating, and perpetuating psychosocial factors associated with FCD may also act as barriers to effective diagnosis and care as the associated psychological states may lead to diminished self-efficacy causing FCD patients to become less likely to engage in needed psychological treatment. Further, clinicians may feel uncomfortable when addressing psychosocial issues, they cannot "solve" which then can lead to patient interactions that overlook the FCD patients' actual emotional context. Thus, identification and greater clinical understanding of all possible psychosocial factors may help to explain why individuals experience distressing cognitive symptoms, like memory loss, word-finding issues, or poor concentration, with or without meeting diagnostic criteria for dementia or its prodromes patterns. This understanding is most certainly critical in the diagnostics and management of mental health conditions presenting within memory or dementia clinics, particularly regarding complex conditions such as FCD.

References

Cabreira V, Frostholm L, McWhirter L, Stone J, Carson A. Clinical signs in functional cognitive disorders: a systematic review and diagnostic meta-analysis. J Psychosom Res. 2023;173:111447.

Cabreira V, McWhirter L, Stone J, Carson A. Precipitating factors for functional cognitive disorder. Eur J Neurol. 2025;32(3):e70082.

Chalak MH, Jafari H, Yadollahifar S, Rajabi F, Zamandi M, Masruri B, Rafati S. The effect of occupational burnout on cognitive failure and general health in industrial workers (case study: Iran Tire Factory). Health Psychol Rep. 2022;11(2):166–75.

du Pont A, Rhee SH, Corley RP, Hewitt JK, Friedman NP. Rumination and executive functions: understanding cognitive vulnerability for psychopathology. J Affect Disord. 2019;256:550–9.

Guarnera J, Yuen E, Macpherson H. The impact of loneliness and social isolation on cognitive aging: a narrative review. J Alzheimers Dis Rep. 2023;7(1):699–714.

Li J, Ortí-Casañ N, Bayraktaroglu I, Mozzanica G, Zhang F, Olivier JDA, Eisel ULM. Psychosocial stressors and cognitive function: an analysis using data from the English longitudinal study of ageing. J Prev Alzheimers Dis. 2025;12(8):100232.

Masurkar AV, Marsh K, Morgan B, Leitner D, Wisniewski T. Factors affecting resilience and prevention of Alzheimer's disease and related dementias. Ann Neurol. 2024;96(4):633–49.

Mavroudis I, Franekova K, Petridis F, Ciobîca A, Dăscălescu G, Anton E, Ilea C, Papagiannopoulos S, Kazis D. Risk, precipitating, and perpetuating factors in functional neurological disorder: a systematic review across clinical subtypes. Brain Sci. 2025a;15(9):907.

Mavroudis I, Franekova K, Petridis F, Ciobîca A, Gabriel D, Anton E, Ilea C, Papagiannopoulos S, Kazis D. Comorbidities across functional neurological disorder subtypes: a comprehensive narrative synthesis. Life. 2025b;15(8):1322.

McEwen BS, Gianaros PJ. Central role of the brain in stress and adaptation: links to socioeconomic status, health, and disease. Ann N Y Acad Sci. 2010;1186:190–222.

Patron VG, Rustomji Y, Yip C, Jenkins LM. Psychiatric comorbidities in functional neurologic symptom disorder. Pract Neurol (Fort Wash Pa). 2022;21(3):71–5. PMID: 36644502; PMCID: PMC9836030.

Pennington C, Ball H, Swirski M. Functional cognitive disorder: diagnostic challenges and future directions. Diagnostics (Basel). 2019;9(4):131.

Picard M, McEwen BS. Psychological stress and mitochondria: a conceptual framework. Psychosom Med. 2018;80(2):126–40.

Rehan S, Phillips NA. Psychosocial function in mild cognitive impairment: social participation is associated with cognitive performance in multiple domains. J Appl Gerontol. 2025;44(10):1629–40.

Shen C, Rolls ET, Cheng W, Kang J, Dong G, Xie C, Zhao X, Sahakian B, Geng J. Associations of social isolation and loneliness with later dementia. Neurology. 2022;99(2):e164–75.

Staunton C, Rudrappa R, Rathnaiah M. The prevalence of attention-deficit hyperactivity disorder in functional neurological disorder: an integrative literature review. BJPsych Open. 2024;10(Suppl 1):S83–4.

Tackey C, Slepian PM, Clarke H, Mittal N. Post-viral pain, fatigue, and sleep disturbance syndromes: current knowledge and future directions. Can J Pain. 2024;7(2):2272999.

Teodoro T, Edwards MJ, Isaacs JD. A unifying theory for cognitive abnormalities in functional neurological disorders, fibromyalgia and chronic fatigue syndrome: systematic review. J Neurol Neurosurg Psychiatry. 2018;89(12):1308–19.

Trumpff C, Monzel AS, Sandi C, Menon V, Klein H, Fujita M, Lee A, Petyuk VA, Hurst C, Duong DM, Seyfried NT, Wingo AP, Wingo TS, Wang Y, Thambisetty M, Ferrucci L, Bennett DA, De Jager P/L, Picard M. Psychosocial experiences are associated with human brain mitochondrial biology. Proc Natl Acad Sci. 2024;121(27):e2317673121.

Vaidya N, Marquand AF, Nees F, et al. The impact of psychosocial adversity on brain and behaviour: an overview of existing knowledge and directions for future research. Mol Psychiatry. 2024;29:3245–67.

Wilson RS, Bennett DA. How does psychosocial behavior contribute to cognitive health in old age? Brain Sci. 2017;7(6):56.

Zahodne LB. Psychosocial protective factors in cognitive aging: a targeted review. Arch Clin Neuropsychol. 2021;36(7):1266–73.

Zannas AS. Gene-environment interactions in late life: linking psychosocial stress with brain aging. Curr Neuropharmacol. 2018;16(3):327–33.

Zhang X, Ge TT, Yin G, Cui R, Zhao G, Yang W. Stress-induced functional alterations in amygdala: implications for neuropsychiatric diseases. Front Neurosci. 2018;12:367.

Sex Also Matters: Menopause, Memory, and Mood

8

Factors for Sex-Specific Cognitive Decline

As discussed in Chap. 3, research on sex-specific cognitive decline reveals a complex "reserve vs. decline" paradox in women versus men regarding cognitive change over time. One such longitudinal study involving a large cohort (i.e., including 26,000 participants) suggest women generally demonstrate higher initial cognitive performance, although they often experience a more rapid rate of decline in cognition in later in life compared to same age men (Levine et al. 2021). More specifically, women were found to typically score significantly higher than men at baseline on measures of global cognition, executive function, and memory. Despite these higher starting cognitive scores, women experience faster annual declines in global cognition and executive function over time. However, the rates of decline in memory-specific measures are often similar between sexes, though some studies suggest women's initial memory advantage might mask underlying pathology until a more sudden decline in cognition occurs. This higher baseline score in women may represent a greater "cognitive reserve," allowing them to maintain normal function longer despite brain pathology followed by more rapid decline when their disease process hits their threshold.

Although the reasons for the sex-specific differences in cognitive decline have not been definitively determined, there are several possible factors that appear to contribute to these findings. This includes both biological and lifestyle factors (Levine et al. 2021) as below:

Biological Risk Factors

- *Tau Pathology*: Among individuals with similar levels of amyloid pathology, women often show higher levels of tau signal in the brain, this "tau burden" is linked to steeper cognitive decline including memory for women (Tsiknia et al. 2022). On neuroimaging studies (PET scans), women show higher levels of

S. Hayden, *Fear of Forgetfulness*, https://doi.org/10.1007/978-3-032-21818-6_8

amyloid at similar stages of Alzheimer's disease. Thus, women show more extensive "downstream" tau damage upon longitudinal cognitive follow up (Zeighami et al. 2025).

- *Hormonal Changes*: The decline of estrogen during menopause is a hypothesized factor in accelerated cognitive aging for women. Estrogen plays a crucial neuroprotective role in brain function, including supporting synaptic plasticity, energy metabolism, blood flow, and protecting neurons from damage. Researchers continue to investigate whether the loss of estrogen at midlife contributes to an increased risk for long-term cognitive decline and neurodegenerative diseases like Alzheimer's disease (Bartkowiak-Wieczorek et al. 2024).
- *APOE ε4 Gene*: While both sexes carry the APOE ε4 gene, its impact on dementia risk and the speed of cognitive decline is reportedly stronger in women. The stronger impact of APOE ε4 gene on women's dementia risk appears related to a complex interplay of hormonal changes, immune system differences, and altered brain metabolism. Estrogen typically provides neuroprotective effects, such as promoting neuronal survival, regulating glucose metabolism, and reducing inflammation. In APOE ε4 carriers, this protective effect is significantly reduced or absent. Studies suggest that while estrogen therapy may benefit non-APOE ε4 women, it can have less favorable, or even detrimental, effects on the cognitive performance of APOE ε4 carriers, depending on the timing and type of therapy (Riedel et al. 2016; Valencia-Olvera et al. 2023).

Socioeconomic and Lifestyle Factors

- *Education*: Lower educational attainment in older women is linked to increased vulnerability for cognitive decline likely as education contributes to cognitive reserve, which helps the brain better cope with age-related changes and potential disease pathology. Notably, historically women are significantly more likely to have lower educational access and attainment than men in aging cohorts which may be contributing to the finding of higher functional limitations reported by women (Bloomberg et al. 2021).
- *Vascular Health*: Factors like high blood pressure and uncontrolled blood sugar appear to have a more pronounced negative impact on the rate of decline in cognition for women than in men. Studies have found that midlife hypertension is an independent risk factor for cognitive decline, but it affects men and women differently. In women, hypertension is associated with worse performance across multiple cognitive domains, including attention, working memory, and associative memory. In men, hypertension is primarily linked only to attentional deficits. Some studies suggest that the effect of high blood pressure on the risk of Alzheimer's disease or vascular dementia may be present in women, but not in men of similar age groups (Thomas et al. 2022).
- *Activity Levels*: Better cognitive function is strongly linked to higher physical activity in women, whereas men's activity levels show less correlation with their cognitive trajectory. A large body of evidence indicates that physical activity has

a more pronounced positive effect on overall cognitive function and a greater ability to mitigate cognitive decline in older women compared to older men. The specific cognitive domains that benefit from physical activity differ by sex, with women often showing greater improvements in executive function and visuospatial abilities while men tend to show greater gains or preservation in memory-related tasks. Further, inactive women experience a steeper cognitive decline than active women and a steeper decline than men, regardless of the men's activity levels, which highlights women's greater sensitivity to activity changes (Li et al. 2025).

Cognitive Function in Women

As mentioned in Chap. 3, overall cognitive function and intellect in men versus women is generally equivalent, albeit with some minor differences in the brain pathways taken to perform various target tasks. Regardless, neuroscience literature (Mitchell and Woods 2001; Ciafaloni 2018) suggests critical periods in women's reproductive life cycle are associated with cognitive issues including the following:

- Cognition fluctuation during menstrual cycle (i.e., such as improved working memory/attention during pre-ovulation when estrogen is highest, with brain fog during days just prior to menstruation).
- Fifty to eighty percent of women report cognitive changes during pregnancy (i.e., issues with attention, verbal fluency, and memory particularly in second and third trimester of pregnancy).
- Sixty-two percent of perimenopausal women report memory change.

These critical periods are characterized by fluctuations or drops in estrogen and progesterone which are associated with various cognitive, emotional, and neurophysiological issues. Understanding these cognitive fluctuations for women is particularly important when considering the possibility of cognitive decline as women age. Women should be encouraged to discuss such issues within clinical settings and clinicians should strive to explore such factors when considering the context of cognitive decline for women.

The overall cognitive impact of women's hormonal fluctuation throughout their cycle, well as throughout their lifespan, is critical to precision medicine and prevention of brain health risk for women. Perhaps one of the most compelling and important research findings in this area is related to the neuroprotective effects of estrogen for women. The impact of such sex hormones on cognition becomes increasingly apparent as they are now believed to play an important role in the protective and antioxidant effects involved in the brain's resilience and neuronal survival (Zárate et al. 2017). It is within this frame of reference that our understanding of increased risk of neurological and mental health conditions seems most prominent for women during transition to menopause and at other times of hormonal fluctuation.

Brann et al. (2022) provide a comprehensive review of the wide-reaching impact of estrogen on brain function including cognition, neuroprotection, and neuroplasticity functions. This understanding of estrogen's integral role, not only in the expression of brain function through cognition but also in protection from brain-related conditions and disease, warrants ongoing investigation. Such research continues as with evidence that estrogen acts as a master regulator for brain health in women, influencing both cognition and mood primarily through estrogen receptors (i.e., the lock that estrogen acts as a key for to trigger specific actions) (Cho et al. 2025). Notably, estrogen has been found to enhance verbal and visual memory as well as aspects of executive function (i.e., mental flexibility) and attention.

Further, estrogen appears to play a neuroprotective role against age-related stressors and neurodegeneration through reduction of neuroinflammation as well contributing to the clearance of amyloid and inhibiting tau hyperphosphorylation. There is a role estrogen also plays in neuroplasticity with involvement in hippocampal neurogenesis, directly related to memory (Sahab-Negah et al. 2020).

Menopause and Cognitive Change

Menopause and the transition to menopause (i.e., perimenopause) are associated with transient changes in cognition and an increased vulnerability to mood disorders, such as depression and anxiety (Fidecicchi et al., 2024). These symptoms are linked to fluctuating and declining estrogen levels, as well as related issues like sleep disturbances and hot flashes. A significant number of women (i.e., 35%–62%) experience cognitive symptoms, commonly self-referred to as "brain fog," during perimenopause and menopause. Notably, although brain fog is considered a real and significant symptom by neuropsychologists and other healthcare professionals, it is not a formal medical diagnosis. Rather, brain fog is a descriptive, subjective term for a collection of very real cognitive challenges. In clinical practice, the more formal terms might be referred to as cognitive dysfunction or cognitive impairment.

Symptoms cited by perimenopausal or menopausal women typically involve subtle difficulties with various aspects of cognitive function that may or may not directly impact daily life and occupation. These changes are primarily linked to the effect of declining estrogen on brain regions critical for memory and executive function, such as the hippocampus and prefrontal cortex. Table 8.1 provides summary of cognitive symptoms most often reported in perimenopause and menopause (Table 8.1):

For most women, these cognitive changes are temporary and tend to normalize after the transition through menopause is complete and they enter the postmenopausal stage of life (i.e., the ongoing time period following menopause defined as 12 months following final menstruation). The cognitive symptoms associated with perimenopause/menopause are generally not an early sign of dementia, which is relatively rare in midlife. The severity of cognitive issues can be exacerbated by other menopausal symptoms such as poor sleep quality, hot flashes, and a depressed mood (Metcalf et al. 2023). With regard to symptom management, estrogen's

Table 8.1 Cognitive symptoms of perimenopause/menopause

Cognitive domain	Functional symptoms
Memory	A common complaint for women in both perimenopause and menopause such as reports of forgetting appointments or losing keys
Attention/ concentration	Issues of focus or sustaining attention when engaging in activities such as reading or watching a movie. This may also entail the "brain fog" reported by women reflecting issues with clear thinking
Language (expressive)	Women may experience problems with finding words in impacting the fluency in conversations during menopause and perimenopause
Processing speed	Women report a general decrease in processing speed and mental slowing at times. This may impact response speed in time sensitive tasks
Executive function	Struggling with managing multiple tasks at one time and shifting from one task to another is a symptom reported by women in and around menopause

cognitive enhancing effects and neuroprotective features have been found to be most effective when hormone therapy (HT) is initiated during the menopausal transition or early postmenopause (Sahab-Negah et al. 2020) with evidence suggesting starting HT over age 65 may not provide these benefits.

Menopause and Mood

The perimenopausal period, referring to the years of transition prior to onset on menopause, acts as a "window of vulnerability" for the development new or recurrent depressive and anxiety symptoms for women (Weber et. al., 2014; Shea et al., 2023). This can include mood swings, irritability, heightened anxiety, sadness, and in some cases, major depressive episode. Neuroscience research (Lang et al. 2025) suggests reasons for such changes may be consequent to estrogen modulating key neurotransmitters like serotonin and dopamine, which are crucial for mood regulation. The significant fluctuations and eventual decline in estrogen can disrupt these systems, leading to mood instability. Further, neuroimaging studies within this research paper summarize evidence for structural brain changes associated with menopausal transition, including reduced volume in emotion-regulating regions such as the hippocampus, amygdala, and prefrontal cortex.

Notably, anxiety is a common and significant complaint during menopause and transition to it. Research reflects a strong and bidirectional correlation with anxiety and the intensity/frequency of vasomotor symptoms such as hot flashes and night sweats (Lucena et al. 2025). Thus, women with more frequent or intense hot flashes would be more likely to experience higher levels of anxiety, which in turn may trigger more intense vasomotor symptoms. This bidirectional relationship has also been found with vasomotor symptoms and depression. These rates of anxiety and depression are often highest in the late stages of perimenopause or the early phases postmenopause, which is also when vasomotor symptoms seem to peak. Women with a history of depression or anxiety, significant life stressors, poor social support, or severe physical symptoms (like frequent hot flashes and sleep problems) are at a higher risk.

Functional Cognitive Disorder (FCD) in Women

As suggested in prior chapters, FCD is more prevalent in women versus men. Although the reasons remain unclear, the higher impact on women is likely not due to one factor but rather an interplay of biology (hormones, genetics), social experiences (trauma), and how the brain ages, making them more susceptible to these functional disorders (McLoughlin et al. 2023). Such factors that appear to play a role in women's risk for FCD are explored in greater detail below:

Hormonal and Biological Factors

Perimenopause and menopause bring significant estrogen fluctuations and ultimately decline, impacting memory and brain function, increasing cognitive complaints and risk for cognitive decline. Biologically, as mentioned in Chap. 3, women appear to be vulnerable to pathological changes like tau tangles, showing symptoms earlier and with increased risk to Alzheimer's disease. Further, women's chromosomal makeup with genes on the X chromosome (i.e., like USP11) may increase vulnerability to tau pathology, a hallmark of neurodegenerative diseases (Conde et al. 2021).

Social and Trauma-Related Factors

Women are disproportionately affected by trauma with higher rates of childhood abuse, intimate partner violence, and sexual assault. Such adverse events, as discussed earlier, represent risk factors for developing mental health issues including functional neurological disorders (FNDs), such as FCD. Also, broader societal factors and gender roles contribute to different exposures and coping mechanisms, influencing risk for women's cognitive decline over time. Interestingly, there is some evidence for a paradoxical effect on dementia risk from advances in gender equality. Improved access to factors such as education, employment, and healthcare appears to increase risk for dementia due to lifestyle factors such as increased smoking and alcohol use by women in such instances (You 2025).

Overall, despite the paucity of research that has been available regarding the higher prevalence of functional neurological disorders and functional cognitive disorders in women, emerging studies are beginning to explore possible reasons for this discrepancy in prevalence of FCD in women. As McLoughlin et al. (2023) note:

> There is abundant evidence that disorders perceived as primarily affecting women are neglected in healthcare and medical research, and the course of FND mirrors this neglect.

We outline the reasons why FND is a feminist issue, incorporating historical and contemporary clinical, research and social perspectives. We call for parity for FND in medical education, research and clinical service development so that people affected by FND can receive the care they need.

References

Bartkowiak-Wieczorek J, Jaros A, Gajdzińska A, Wojtyła-Buciora P, Szymański I, Szymaniak J, Janusz W, Walczak I, Jonaszka G, Bienert A. The dual faces of oestrogen: the impact of exogenous oestrogen on the physiological and pathophysiological functions of tissues and organs. Int J Mol Sci. 2024;25(15):8167.

Bloomberg M, Dugravot A, Dumurgier J, Kivimaki M, Fayosse A, Steptoe A, Britton A, Singh-Manoux A, Sabia S. Sex differences and the role of education in cognitive ageing: analysis of two UK-based prospective cohort studies. Lancet Public Health. 2021;6(2):e106–15.

Brann D, Lu Y, Wang J, Zhang Q, Thakkar R, Sareddy GR, Pratap U, Tekmal RR, Vadlamudi R. Brain-derived estrogen and neural function. Neurosci Biobehav Rev. 2022;132:793–817.

Cho JM, Lee J, Ahn EM, Bae J. Beyond hot flashes: the role of estrogen receptors in menopausal mental health and cognitive decline. Brain Sci. 2025;15(9):1003. https://doi.org/10.3390/brainsci15091003.

Ciafaloni E. Cognitive changes in pregnancy. In: Ciafaloni E, Thornburg LL, Bushnell CD, editors. Neurological diseases and pregnancy: a coordinated care model for best management. New York: Oxford Academic; 2018.

Conde DM, Verdade RC, Valadares ALR, Mella LFB, Pedro AO, Costa-Paiva L. Menopause and cognitive impairment: a narrative review of current knowledge. World J Psychiatry. 2021;11(8):412–28.

Fidecicchi T, Giannini A, Chedraui P, Luisi S, Battipaglia C, Genazzani A, Genazzani A, Tommaso Simoncini T. Neuroendocrine mechanisms of mood disorders during menopause transition: a narrative review and future perspectives. Maturitas. 2024;188:108087.

Lang XL, Huang CC, Cui HY, Zhong HX, Shen MY, Zhao F. From physiology to psychology: an integrative review of menopausal syndrome. World J Psychiatry. 2025;15(11):108713.

Levine DA, Gross AL, Briceño EM. Sex differences in cognitive decline among US adults. JAMA Netw Open. 2021;4(2):e210169.

Li W, Chen P, Jiang A, Zhang F, Wu Z. Gender differences in cognitive benefits of meeting physical activity guidelines in older Chinese adults. Front Public Health. 2025;13:1539369.

Lucena L, de Campos BH, Dimov L, Tufik S, Hachul H. Is anxiety associated with vasomotor symptoms in postmenopausal women with insomnia complaints? Sleep Sci. 2025;18(4):e410–9.

McLoughlin C, Hoeritzauer I, Cabreira V, Aybek S, Adams C, Alty J, Ball HA, Baker J, Bullock K, Burness C, Dworetzky BA, Finkelstein S, Garcin B, Gelauff J, Goldstein LH, Jordbru A, Huys AM, Laffan A, Lidstone SC, Linden SC, Ludwig L, Maggio J, Morgante F, Mallam E, Nicholson C, O'Neal M, O'Sullivan S, Pareés I, Petrochilos P, Pick S, Phillips W, Roelofs K, Newby R, Stanton B, Gray C, Joyce EM, Tijssen MA, Chalder T, McCormick M, Gardiner P, Bègue I, Tuttle MC, Williams I, McRae S, Voon V, McWhirter L. Functional neurological disorder is a feminist issue. J Neurol Neurosurg Psychiatry. 2023;94(10):855–62.

Metcalf CA, Duffy KA, Page CE, Novick AM. Cognitive problems in perimenopause: a review of recent evidence. Curr Psychiatry Rep. 2023;25(10):501–11.

Mitchell SE, Woods FN. Midlife women's attributions about perceived memory changes: observations from the Seattle Midlife Women's Health Study. J Women Health Gend Based Med. 2001;10:351–62.

Riedel BC, Thompson PM, Brinton RD. Age, APOE and sex: triad of risk of Alzheimer's disease. J Steroid Biochem Mol Biol. 2016;160:134–47.

Sahab-Negah S, Hajali V, Moradi HR, Gorji A. The impact of estradiol on neurogenesis and cognitive functions in Alzheimer's disease. Cell Mol Neurobiol. 2020;40(3):283–99.

Shea AK, Wolfman W, Fortier M, Soares CN. Menopause: mood, sleep, and cognition. J Obstet Gynaecol Can. 2023;43(11):1316–1323.e1.

Sherwin B. Estrogen and cognitive function in women. Endocr Rev. 2003;24(2):133–51.

Thomas EG, Rhodius-Meester H, Exalto L, Peters SAE, van Bloemendaal L, Ponds R, Muller M. Sex-specific associations of diabetes with brain structure and function in a geriatric population. Front Aging Neurosci. 2022;14:885787.

Tsiknia AA, Edland SD, Sundermann EE, Reas ET, Brewer JB, Galasko D, Banks SJ. Alzheimer's Disease Neuroimaging Initiative. Sex differences in plasma p-tau181 associations with Alzheimer's disease biomarkers, cognitive decline, and clinical progression. Mol Psychiatry. 2022;27(10):4314–22.

Valencia-Olvera AC, Maldonado Weng J, Christensen A, LaDu MJ, Pike CJ. Role of estrogen in women's Alzheimer's disease risk as modified by APOE. J Neuroendocrinol. 2023;35(2):e13209.

Weber MT, Maki PM, McDermott MP. Cognition and mood in perimenopause: a systematic review and meta-analysis. J Steroid Biochem Mol Biol. 2014;142:90–8.

You W. Dementia risk and gender equality: global insights into social determinants. Nurs Health Sci. 2025;27(3):e70230.

Zárate S, Stevnsner T, Gredilla R. Role of estrogen and other sex hormones in brain aging. Neuroprotection and DNA repair. Front Aging Neurosci. 2017;9:430.

Zeighami Y, Tremblay C, Dadar M. Sex differences in vulnerability to tau pathology: impact on cognitive decline. medRxiv [Preprint]. 2025:2025.05.25.25328324.

Part III

Applied Neuroscience in Brain Health & Dementia

9 Applied Neuroscience: The Path to Prevention and Symptom Management

Applied neuroscience, in general terms, refers the practical, real-world application of brain research to improve how we live, work, and learn. This area of neuroscience may be more specifically applied to understanding, not only the cause of disease, but also the prevention and management of neuropsychiatric and neurodegenerative disorders (Meyer-Lindenberg et al. 2022). With respect to working within the context of memory or dementia clinics, applied neuroscience relies upon research to improve how we diagnose, manage, and understand cognitive function and disorders. This area of neuroscience engages tools such as advanced neuroimaging and neuropsychological testing to determine whether cognitive issues stem from structural brain damage as in the various forms of dementia or may be related to potentially reversible disruption in brain function as in functional cognitive disorders (FCD). Clearly, such tools are useful not only in the identification and diagnosis of cognitive disorders, but may also be critical to the prevention and management of cognitive symptoms for patients presenting to memory or dementia clinics. As the role of such tools in diagnostics has been discussed within prior chapters, the focus here will be upon the prevention and management aspects of applied neuroscience for cognitive disorders presenting to memory or dementia clinics (Guzman-Martinez et. al., 2021).

Path to Prevention

An applied neuroscience-informed prevention path for cognitive disorders includes strategies targeted at underlying risk factors that include leveraging the brain's inherent capacity for change, known as *neuroplasticity*. The most basic definition of neuroplasticity refers to the "brain's capacity to reorganize itself by forming new neural connections" (Gazerani 2025). While neuroplasticity is the brain's ability to form new neural connections, *neurogenesis* refers to the brains capacity to generate of new neurons (Vaz et al. 2022). Neurogenesis is thought to be a structural aspect of neuroplasticity that happens primarily in the hippocampus which allows the brain

S. Hayden, *Fear of Forgetfulness*, https://doi.org/10.1007/978-3-032-21818-6_9

to learn and form new memories. Both neurogenesis and neuroplasticity play a role in prevention of cognitive disorders, including both dementia and FCD.

Providing psychoeducation regarding neuroplasticity is an important component to clinical communications on prevention of cognitive disorders. A common analogy for neuroplasticity that I often use with my own patients, and perhaps most relevant as a Canadian who is well familiar with snow, can enhance patient understanding of its importance for "beefing up" brain tissue. The description below is one I commonly use in clinical practice when discussing neuroplasticity with patients in our memory clinic at UBC hospital:

> *So, neuroplasticity is your brain's ability to form new neural connections through mental exercise and activities that theoretically strengthen brain tissue by creating novel pathways. One way to conceptualize this is through an analogy of walking through a field of deep fresh snow:*
>
> - *The Initial Walk (Learning/New Experiences) When you walk through deep, fresh snow for the first time, it requires significant energy, effort, and concentration. Every step is slow and difficult. This represents the brain learning a new skill and forming a new neural pathway.*
> - *Repetition (Strengthening Pathways)*: Now if you walk that same path a second or third time, the walk becomes easier. The snow becomes packed down as you walk more, and a clearer trail emerges. In the brain, repeating a behavior or thought strengthens the connection between neurons, making that pathway more efficient and stronger.
> - *Changing Patterns (Rewiring the Brain)*: Neuroplasticity allows you to create new, different paths in the snow. While it is difficult to break away from the established "highway" and trudge through fresh snow to create a new route, doing so through intentional, repeated action allows you to build new, healthier neural pathways. This learning of new skills may provide some protection from neurodegenerative disease, such as Alzheimer's disease.

Although the capacity for neuroplasticity had historically been thought to be present only in developing brains (i.e., in childhood through early 20s), there is now recognition that neuroplasticity exists throughout our lifespan. Such concepts were perhaps first explored within the "nun study" of 1986, where researchers examined a large group of nuns with similar lifestyle approaches and risk for dementia. In the initial study, Dr. Snowdon examined lifestyle factors and cognitive function in the nuns of the School Sisters of Notre Dame (SSND), who had engaged in lifelong learning as a component of their convent's culture. This body of research led to discoveries that some nuns who were cognitively intact during their lifetime (i.e., demonstrated within repeat neuropsychological evaluations for the study) were later found to have significant brain pathology for Alzheimer's disease (i.e., plaques and tangles) upon autopsy. This led to the suggestion that higher education and lifelong learning may build a cognitive reserve that acts as protection from expression of dementia symptoms (Clarke et al. 2025).

Subsequently, researchers have come to believe neuroplasticity may have both positive and negative consequences. The *positive* side of neuroplasticity includes adaptation/rehabilitation, skill acquisition, and healthy aging related to the development of new neural pathways as described earlier. The *negative* side of neuroplasticity entails maladaptive entrenchment of addiction, pain, and psychiatric issues (Marzola et al. 2023), including symptoms of conditions such as functional cognitive disorder (FCD). These negative consequences are hypothesized to relate to a self-reinforcing downward spiral (Tomaszczyk et al. 2014), including the vicious cycle of symptoms and cognitive dysfunction illustrated within prior chapters of this book.

Applied neuroscience research has explored means of intervention to mitigate such negative consequences to neuroplasticity, including diminishment of neurogenesis as well as maladaptive neuroplasticity. Gazerani (2025) provides a review of current evidence regarding such studies, revealing moderate levels of efficacy for various types of interventions, including lifestyle interventions such as diet, exercise, and sleep optimization, when applied to manage symptoms of various neuropsychiatric and neurologic conditions. Such research has helped to identify genetic, environmental, and lifestyle factors that may contribute to the development of neurological and psychiatric disorders. For instance, such studies analyze the interaction between diet, physical exercise, sleep regulation, and social interactions in maintaining brain health and preventing cognitive decline in later life.

Generally, non-pharmacological approaches, including lifestyle modifications, found to be of benefit within applied neuroscience research are crucial tools in the prevention of cognitive decline. Although some of these interventions will be reviewed in greater detail within the next chapter, an introduction to factors that impact cognitive health includes the following:

- *Physical and Mental Exercise*: Both physical exercise and mentally stimulating activities (i.e., learning novel skills or engaging in mental exercises) have been found to be essential for sustaining brain health and potentially preventing conditions like Alzheimer's disease (AD). In fact, there appears to be emerging neuroscience research to suggest a *synergistic effect* (i.e., where using a combination of factors has greater effect) with use of physical and mental exercise. This study appears to indicate that physical exercise and cognitive training activities involve different mechanisms of brain plasticity and thus have greater positive impact for prevention when combined (Bherer et al. 2021). These researchers note that physical exercise appears to have greater impact on neurogenesis, while cognitive exercise seems to be associated with greater brain connectivity and higher brain volume. The importance of physical inactivity as a dementia risk factor was further supported in a study (Son et. al., 2024), which identified 12 potentially modifiable dementia risk factors accounting for approximately 50% of dementia cases in Canada.
- *Neuromodulation*: Techniques like deep Transcranial Magnetic Stimulation (dTMS) are being explored for the treatment of mental health and cognitive disorders. This is a noninvasive brain stimulation technique using a fitted helmet to send magnetic pulses to deeper brain regions to activate neural networks involved in mood and cognition. There is some evidence of benefit to treatment-resistant

depression and OCD (Cheng et al. 2023), which as stated in previous chapters result in cognitive symptoms for patients attending memory clinics.

- *Mind-Body Practices*: Interventions such as meditation, yoga, and stress reduction techniques help develop health-promoting behaviors and appear to foster emotional and cognitive resilience to deterioration with aging. Emerging evidence suggests such "mind-body" interventions may have direct impact on brain structures and function to improve both cognitive and emotional resilience (Esch and Stefano 2022). Such mind-body practices, therefore, may then act as effective preventative tools for cognitive disorders including dementia and possibly FCD (Marín-Medina et. al., 2025).
- *Nutritional Neuroscience*: An emerging field of applied neuroscience is exploring the role of diet, macro- and micronutrients, and gut microbiota in brain health (Ajay et al. 2026). Such research suggests that the gut microbiome may significantly influence brain function via the gut-brain axis. Thus, gut microbiota appears to impact mood, cognition, and behavior and warrants ongoing research to explore more specific clinical indications for individuals presenting to memory or dementia clinics.
- *Cognitive Behavioral Therapy (CBT)*: Applied neuroscience has provided a framework for understanding how highly effective, evidence-based psychological treatments like CBT work by targeting specific cognitive processes that maintain disorders, as discussed in prior chapters. This form of psychological treatment is particularly important for patients presenting with FCD symptoms within the context of investigations for neurodegenerative disorders in memory clinics.

Applied Neuroscience for Functional Cognitive Disorders

Applied neuroscience also informs the understanding and treatment of functional cognitive disorders (FCDs) by investigating how brain network dysfunction, rather than structural damage, leads to symptoms. This perspective helps guide therapies focused on retraining brain responses and improving self-monitoring abilities. Overall, research into the neurobiological theories of FCD suggests the following regarding the underlying causes:

- *Brain Network Dysfunction*: Applied neuroscience research suggests that FCD involves alterations in how brain networks, such as those related to attention, self-agency, and emotional processing, communicate with each other. So, based on recent neuroimaging studies, functional cognitive disorder (FCD) is increasingly understood as a disorder of brain network dysfunction rather than structural damage (Jungilligens et al. 2022).
- *Abnormal Predictive Coding*: One leading model posits a mismatch between "top-down" predictions (i.e., expectations about one's cognitive performance) and "bottom-up" sensory information (i.e., input from the body or environment). Increased attention to symptoms can amplify this disconnect, perpetuating the

disorder (Fiorio et al. 2022). As discussed in prior chapters, this leads to the vicious cycle of cognitive hypervigilance and high cognitive expectations leading to increased cognitive errors on a functional basis.
- *Impaired Metacognition*: Patients with FCD often experience difficulties with metacognition, which refers to the ability to accurately self-evaluate one's own cognitive performance. This can manifest as "memory perfectionism" or heightened self-monitoring in those with FCD, where minor lapses in attention (common in healthy individuals) are perceived as significant and distressing, reinforcing the belief in a severe problem (Franekova et al. 2025) and leading to the misinterpretation that such cognitive issues represent an emerging dementia. It is often this issue that is prevalent in those FCD patients presenting to memory clinics.

The proposed mechanism of what underlies FCD is further advanced by suggestion that there is a wide range of behavioral (i.e., including perfectionism and increased self-monitoring of cognitive processes), biological, and environmental factors that may then lead to attention lapses and overinterpretation of these lapses in daily life. This, in turn, can trigger the prediction of future lapses, leading to a vicious cycle of inefficient cognitive function on a functional basis as mentioned in previous chapters.

Applied Neuroscience in Treatment

The neuroscience-based pathophysiology, which involves attentional dysregulation, impaired metacognition, and limbic overactivity rather than structural damage, directly informs specific, evidence-based treatments such as psychoeducation, Cognitive Behavioral Therapy (CBT), and metacognitive retraining. Thus, neuroscience of FCD translates into specific, evidence-based treatments such as below:

- *Psychoeducation and Diagnosis Explanation*: As mentioned previously, providing a clear, positive diagnosis and explaining the neurological basis of the symptoms helps alleviate patient uncertainty and anxiety which can then mitigate FCD symptom triggers and maintenance. Understanding that the symptoms of FCD arise from a "software" issue in brain function rather than a "hardware" problem (like a structural disease as in dementia) is a crucial first step towards recovery for patients in this category. This recovery process begins with providing psychoeducation on the difference between *hardware* (i.e., structural disease such as brain damage, tumors, or neurodegeneration/Alzheimer's disease) and *software* (i.e., functional cognitive disorder). With software issues, the brain is structurally healthy, but the network-level functioning, specifically how it processes information, sends, and receives signals, is malfunctioning. This framing helps patients understand that the brain is not degenerating, but rather, the brain's software needs to be reprogrammed to improve functioning. Generally, such an approach confirms that the patient's symptoms are real, not "imagined" or "all in the mind," but rather a genuine disorder of brain functioning. Communicating the process of FCD in this manner, using knowledge acquired through applied neuroscience, limits the possibility of self-stigmatization and reduces patients' feeling dismissed within clinical settings. Further, clear communication that FCD is non-degenerative helps ease the fear of a dementia diagnosis. It shifts the focus from hunting for a disease (which causes anxiety) to rehabilitation and management of the "software" issue. Such a shift in focus serves to empower patients in a positive manner, which is particularly important for those presenting with cognitive concerns related to mental health factors such as FCD.

- *Cognitive Behavioral Therapy (CBT)*: CBT is used to manage predisposing or perpetuating factors like anxiety, depression, and stress, which can exacerbate symptoms. It also helps patients develop tools to cope with rigid thinking and excessive worry about cognitive lapses. As mentioned, CBT may include exposure therapy for those FCD patients requiring such intervention in a safe and controlled manner. Although CBT techniques will not be reviewed in detail here, the efficacy of such treatment for a vast array of conditions is well established within research including applied neuroscience.
- *Cognitive Rehabilitation*: This approach uses techniques and strategies to improve attention, memory, and executive function Although such techniques cannot substantively reverse cognitive decline for patients with dementia, such interventions can be particularly effective for patients with cognitive concerns related to mental health factors such as FCD, particularly if combined with other treatment modalities (i.e., such as CBT). Furthermore, such techniques and strategies can provide a sense of empowerment for those struggling with cognitive concerns, particularly in the face of fear for forgetfulness, whether these cognitive concerns are related to an underlying mental health condition or to an emerging neurodegenerative condition. Cognitive rehabilitation may involve the following:
  - *Metacognitive Training*: Computer-based or other training to improve self-monitoring and self-evaluation of cognitive abilities.
  - *Mnemonic Techniques*: Learning strategies to structure information recall and use external memory aids.
  - *Attention Modification*: Techniques to optimize attention and reduce hypervigilance toward internal symptoms.
- *Addressing Comorbidities*: Treatments for FCD or other mental health disorders with associated cognitive concerns also often include addressing related conditions such as sleep disorders, chronic pain, or physical fitness, as these factors can influence cognitive function and mental health. Notably, addressing such comorbidities is also critical for those experiencing symptoms associated with an emerging neurodegenerative condition.
- *Physical and Occupational Therapy*: Though often associated with motor symptoms of functional neurological disorder (FND), therapists such as PTs and OTs can help in a general capacity by optimizing physical and functional capacities, which in turn supports overall brain health and function. Such interventions may

also support the preventative measures for brain health that require physical activation.

Given the complexity of cognitive disorders related to or exacerbated by mental health factors, including FCD, particularly within the context of investigations for neurodegenerative diseases, it is likely that the most effective treatment/management of symptoms related to this syndrome requires a multifaceted approach rather than just one of the above approaches (Barbey et al. 2022). Thus, no single tool is likely to prove effective to prevent or mitigate cognitive symptoms for those presenting to memory clinics.

The Future of Applied Neuroscience

The field of Applied Neuroscience continues to evolve, integrating basic research with clinical applications to improve health outcomes. A major goal is to use neuroscientific tools to enhance diagnostic specificity, identify vulnerability in at-risk populations, and improve intervention efficacy. This promises a future with more personalized and effective ways of understanding and managing brain and mental health. Such research is particularly critical to understanding the complexities of mental health factors, such as functional cognitive disorder (FCD), in the presentation of cognitive disorders to memory clinics. As psychoeducation regarding the cause of functional cognitive errors is a critical step in the management of such patients presenting with fear for cognitive decline to memory clinics, advancements in applied neuroscience are key to facilitating clinican-patient communications regarding accurate diagnosis and mitigation of cognitive and/or mental health symptoms.

References

Ajay A, Singh Gaur S, Singh J, Shams R, Dash K, Mukarram A, Kovács B. Dietary influences on the gut-brain pathways: mechanisms and therapeutic potential. Hum Nutr Metab. 2026;43:200349.

Barbey A, Pjanic I, Studer H, Bischoff N, Bassetti CLA, Aybek S. Management of functional neurological disorders (FND): experience from a Swiss FND clinic. Clin Transl Neurosci. 2022;6(1):2.

Bherer L, Gagnon C, Langeard A, Lussier M, Desjardins-Crépeau L, Berryman N, Bosquet L, Vu TTM, Fraser S, Li KZH, Kramer AF. Synergistic effects of cognitive training and physical exercise on dual-task performance in older adults. J Gerontol B Psychol Sci Soc Sci. 2021;76(8):1533–41.

Cheng JL, Tan C, Liu HY, Han DM, Liu ZC. Past, present, and future of deep transcranial magnetic stimulation: a review in psychiatric and neurological disorders. World J Psychiatry. 2023;13(9):607–19.

Clarke KM, Etemadmoghadam S, Danner B, Corbett C, Ghaseminejad-Bandpey A, Dopler M, Parker-Garza J, Alhneif M, Babu S, Ogunbona OB, Gonzalez AD, Salardini A, Flanagan ME. The Nun Study: insights from 30 years of aging and dementia research. Alzheimers Dement. 2025;21(2):e14626.

Esch T, Stefano GB. The BERN framework of mind-body medicine: integrating self-care, health promotion, resilience, and applied neuroscience. Front Integr Neurosci. 2022;16:913573.

Fiorio M, Braga M, Marotta A, Villa-Sánchez B, Edwards M, Tinazzi M, Barbiani D. Functional neurological disorder and placebo and nocebo effects: shared mechanisms. Nat Rev Neurol. 2022;18:1–12.

Franekova K, Petridis F, Ciobica A, et al. Metacognitive performance in functional cognitive disorder (FCD): a meta-analysis. Acta Neurol Belg. 2025. https://doi.org/10.1007/s13760-025-02969-8.

Gazerani P. The neuroplastic brain: current breakthroughs and emerging frontiers. Brain Res. 2025;1858:149643.

Guzman-Martinez L, Calfío C, Farias GA, Vilches C, Prieto R, Maccioni RB. New frontiers in the prevention, diagnosis, and treatment of Alzheimer's disease. J Alzheimers Dis. 2021;82(s1):S51–63.

Jungilligens J, Paredes-Echeverri S, Popkirov S, Barrett LF, Perez DL. A new science of emotion: implications for functional neurological disorder. Brain. 2022;145(8):2648–63.

Marín-Medina DS, Arenas-Vargas PA, del Pilar Velásquez-Duque M, Bagnati PM. Functional cognitive disorder: beyond pseudodementia. Neurol Perspect. 2025;5(1):100179.

Marzola P, Melzer T, Pavesi E, Gil-Mohapel J, Brocardo PS. Exploring the role of neuroplasticity in development, aging, and neurodegeneration. Brain Sci. 2023;13(12):1610.

Meyer-Lindenberg A, Dickson SL, Millan MJ, Knudsen GM. Prospects for improved prevention and treatment of neuropsychiatric disorders: Neuroscience Applied. Neurosci Appl. 2022;1:100103.

Son S, Speechley M, Zou GY, Kivipelto M, Mangialasche F, Feldman H, Chertkow H, Belleville S, Nygaard H, Hachinski V, Pieruccini-Faria F, Montero-Odasso M. Potentially modifiable dementia risk factors in Canada: an analysis of Canadian Longitudinal Study on Aging with a multi-country comparison. J Prev Alzheimers Dis. 2024;11(5):1490–9.

Tomaszczyk JC, Green NL, Frasca D, Colella B, Turner GR, Christensen BK, Green RE. Negative neuroplasticity in chronic traumatic brain injury and implications for neurorehabilitation. Neuropsychol Rev. 2014;24(4):409–27.

Vaz A, Ribeiro I, Pinto L. Frontiers in neurogenesis. Cells. 2022;11(22):3567.

Preventative Brain Health 10

Although the concept of prevention for cognitive decline arises from previously discussed studies associated with applied neuroscience, emerging research on the modifiable risk factors for dementia is advancing this area further. Perhaps the most identifiable study was published by the 2020 Lancet Commission (Livingston et al. 2020) where researchers identified key modifiable dementia risk factors, including lower educational levels, hypertension, traumatic brain injury, alcohol, obesity, smoking, depression, social isolation, physical inactivity, diabetes, and air pollution, emphasizing intervention across the lifespan. Subsequently, a 2024 update added vision loss and high cholesterol (Livingston et al. 2024). Addressing these factors, especially through healthy lifestyles (i.e., diet, exercise, not smoking/drinking), is now thought to significantly delay or prevent dementia. These studies have provided important graphic representation of modifiable risk factors, as reflected below, and may be useful for clinicans for psychoeducation use with patient populations.

Within my clinical practice, the finding that 40% or more of dementia cases worldwide may be attributable to modifiable or lifestyle factors has brought a facet of hope for patients and their families as an identifiable means of coping with fear of forgetfulness. Routinely, clinical care within dementia clinics is now incorporating preventative brain health practices which serve to empower patients experiencing "fear of forgetfulness" regardless of the cause for their cognitive and functional struggles (Fig. 10.1).

In order to further understand the importance of prevention, one must explore the role of lifespan risk in development of diseases such as dementia. As can be seen in the above illustration, various risk factors present at different ages within the lifespan can impact dementia development. For instance, higher education in youth may build cognitive reserve and optimize neuroplasticity as a preventative factor in future development of dementia. In midlife, health status such as blood pressure can represent a mitigating risk factor for future cognitive decline and in late-life factors like inactivity appear to matter. Further, it is important to acknowledge that having multiple risk factors, such as those illustrated, may increase risk more significantly

S. Hayden, *Fear of Forgetfulness*, https://doi.org/10.1007/978-3-032-21818-6_10

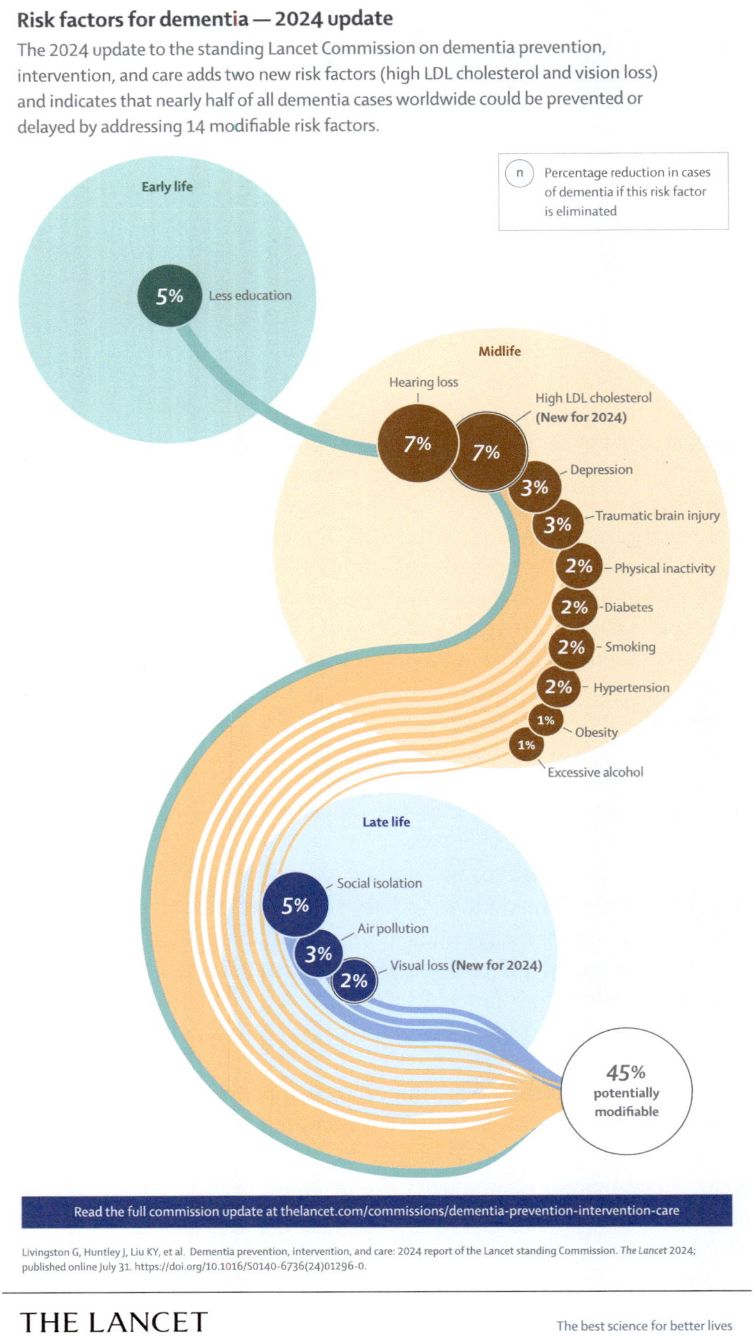

Fig. 10.1 Cognitive and functional struggles (Livingston et al. 2024)

than a single factor alone for an individual's future cognitive decline. Consequently, a multifaceted approach to prevention to mitigate such risks is critical when considering means for addressing cognitive decline and/or the fear of such decline, particularly in patients experiencing cognitive issues related to mental health such as in functional cognitive disorder (FCD). Focussing upon scientifically supported factors thought to optimize brain health can empower patients with cognitive concerns presenting to memory or dementia clinics, regardless of the cause (Chapman et al. 2021). Clinician-patient communications may be facilitated by this science whether a patient is presenting concerns related to neurodegenerative processes or to mental health factors, such as FCD, or to any co-morbidities.

Core Lifestyle Habits for Brain Health

Although there are multiple models for brain health prevention, most models include similar basic components (Potashkin et al. 2025), based upon applied neuroscience and dementia prevention research. Components that are most commonly included within models of a multifaceted brain health are illustrated below (Fig. 10.2).

The multidimensional nature of brain health prevention is evident within this illustration and has been well supported within scientific literature. As with synergistic benefit of physical and mental exercise mentioned within the prior chapter, likely the optimal approach to brain health is indeed multifactorial rather than based on any one preventative factor. Nevertheless, we will review each of these brain health prevention factors independently below, with the understanding that no single factor is solely preventative for cognitive decline and/or emotional resilience.

Fig. 10.2 Model for brain health

Regular Physical Activity: Engaging in moderate-to-vigorous exercise, such as brisk walking, swimming, or dancing, helps pump blood to the brain, which nourishes cells with oxygen and nutrients, improves mood by releasing endorphins, and can reduce the risk of cognitive decline and dementia. Regular physical activity may lower risk of developing dementia with some studies suggesting it may slow cognitive decline by up to 45% (Pahlavani 2023). The underlying mechanism appears related to neurogenesis in that exercise stimulates the release of growth factors (i.e., such as BDNF) which encourages the survival of brain cells and the development of new neural connections, particularly in the hippocampus an area critical for short-term memory. Exercise has also been found to have benefit or elevate mood through the release of endorphins and serotonin which are hormones used to moderate stress, anxiety, and depression. With respect to recommended dose for physical activity, studies suggest individuals aim for at least 150 minutes of moderate activity per week, or 75 minutes of vigorous activity.

Healthy Nutrition: A balanced diet is another crucial element to preventing cognitive decline and mitigating mental health factors. Diets such as the Mediterranean and MIND diets, which emphasize fruits, vegetables, whole grains, nuts, and fish while limiting processed foods, red meat, and sweets, are associated with a reduced risk of cognitive decline. The omega-3 fatty acids, found in fatty fish, which are prominent in these diets, are particularly important for brain cell function and mood regulation.

Seabrook and associates (2025) provided a comprehensive review of dietary recommendations for brain health across various common dietary patterns. Of note, no specific diet has definitively been determined to be preventative of cognitive decline with specificity, although the three dietary patterns discussed in this paper (i.e., MIND, Mediterranean Diet and DASH) that have most benefit in this regard and are detailed further below.

- *Mediterranean Diet*: Encourages the high intake of fruits, vegetables, whole grains, legumes, nuts, and olive oil, with moderate consumption of fish, poultry, and dairy. There is lower intake of red meat and sugar/desserts. There is allowance for moderate wine consumption. The brain health benefits of the Mediterranean diet include its richness of antioxidants, polyunsaturated fatty acids (PUFAs), and anti-inflammatory compounds. These are associated reduced risks of cognitive decline with the mediterranean diet, including risk reduction for Alzheimer's disease, and other dementias.
- *DASH Diet:* Emphasizes the intake of fruits, vegetables, whole grains, low-fat dairy, lean protein, with reduced sodium intake. The brain health benefit of the DASH diet is primarily lowered blood pressure. Thus, the DASH indirectly supports brain health by improving vascular health, which plays a role in the reduction for risk of stroke and vascular dementia.
- *MIND Diet:* Combines elements of the Mediterranean Diet and DASH as above, along with a focus on specific foods linked to cognitive health, including berries and green leafy vegetables. The brain health benefits of the MIND

diet is its association with slower cognitive decline and lower Alzheimer's risk for similar reasons to the Mediterranean and DASH diets.

Quality Sleep: Adequate, quality sleep (i.e., 7–9 hours per night for most adults) is fundamental for processing memories, clearing toxins (i.e., including amyloid) from the brain, and regulating mood (Eugene & Masiak 2015). Poor sleep is a risk factor for cognitive decline as well as heightened risk for mental health disturbance, and thus it is important to maintain a consistent sleep schedule and create a restful environment (Spira et al. 2014). Such research suggests inadequate sleep, meaning less than 7 hours per night, is associated with higher risk for cognitive impairment and increased incidence of neurodegenerative disorders. Furthermore, poor sleep certainly has a negative impact on cognition by reducing attention/concentration and memory on a functional basis which can then lead to the concern for cognitive decline related to mental health issues such as FCD.

Mental Stimulation: Challenging your brain with new and complex tasks helps build cognitive reserve and promotes new connections between nerve cells, as discussed earlier regarding neuroplasticity. Challenging your brain with new skills, puzzles, and learning keeps it healthy by building new neural connections, which helps maintain function and may play a role in mitigating cognitive decline. However, as with other aspects of brain health, it must be paired with overall healthy habits such as a healthy diet, exercise, sleep, and social interaction for best results. Engaging in unfamiliar, complex tasks, rather than just routine puzzles, fosters significant growth, according to research. Notably, passive activities like excessive TV watching do not offer the same stimulation. There is some evidence for positive impact of brain games on cognitive performances (Al-Thaqib et al. 2018).

Social Engagement: Maintaining strong social ties and interacting with others can help ward off isolation and loneliness, reduce stress, and keep your mind active. More recently, studies suggest that being married or in a relationship, living with one or more person, and never feeling lonely were associated with slower annual decline in global cognition (Samtani et al. 2022). As mentioned in prior chapters, psychosocial interventions aimed at mitigating social isolation and enhance social engagement is an important preventative tool for cognitive issues related to either mental health or neurodegenerative processes.

Stress Management: Chronic stress may directly adversely impact the brain over time. Techniques such as mindfulness, meditation, deep breathing exercises, or yoga can help manage stress and promote emotional resilience. Such stress management techniques appear to protect brain health by activating relaxation responses, reducing inflammation, and counteracting chronic stress's damaging impact upon cognition such as effects on memory and attention (Graham-Engeland et al. 2025). Some of the vast research associated with such mind-body techniques aimed at stress management will be discussed in greater detail below.

Manage Chronic Health Conditions: Conditions such as high blood pressure, high cholesterol, and diabetes are risk factors for dementia. Thus, brain health includes working with healthcare providers to manage these conditions effectively through lifestyle changes and/or medication. As mentioned within the previous chapter, effective management of health issues is an important preventative factor for cognitive concerns related either to possible neurodegenerative disorders or mental health factors, such as FCD.

By adopting these proactive lifestyle and preventative measures, brain health may be more robustly supported including emotional resilience, resulting in potential reduction in the risk of neurodegenerative diseases. Encouraging such measures in the management of patients with cognitive symptoms related to mental health factors is also an important component of their clinical recovery and/or stabilization.

Mind-Body Practices as Prevention

Although many brain health models for prevention categorize mind-body practice under the "stress management," I believe this is an area that warrants special attention, perhaps based upon my own practice of yoga over the last 45 years. Encouragingly, the emergence of mind-body medicine seems to signify a shift in the reductionism of western medicine, albeit at a painfully slow pace. The dogged pursuit of scientific rigor, since the nineteenth century, resulted in medical studies and clinical practice moving toward medical specialization, often reducing a patient's experience to biological components of function (i.e., such as with biomarkers). In recent years, a systems-focused perspective (i.e., focus on the whole picture rather than just a single element) in clinical care has emerged, while still acknowledging the need for reductionism/specialization in medical science. It is from the holism of systems-focused perspective that mind-body medicine emerges as an integral component in providing a well-rounded approach to clinical care.

Before diving into the current scientific literature on mind-body medicine, it is necessary to review of the basics of mind-body practices. In general, mind-body practices combine one or more of the following:

- Mental focus—entails the ability to directly control attention towards a specific task
- Controlled breathing—involves the conscious regulation of breath rate and depth to manage mental states
- Movement—some practices include movement of various forms

Notably, mind-body practices may be easily integrated into conventional medical treatment protocols and easily encouraged by clinicans. Trained teacher/practitioners often teach or administer mind-body practices to others, but they can also be practiced on your own. Some examples of mind-body practices are included in Table 10.1.

Table 10.1 Types of mind-body practices

Meditation	Entails various methods of encouraging awareness and focused attention often using breath work and/or reflection
Yoga	Involves various styles of combining movement with breath and/or relaxation
Tai Chi	Is a noncombative martial art with a series of slow controlled movement
Mindfulness	Is a practice of purposeful nonjudgmental focus on the present
Qi Gong	Is a system of coordinated movement, breath, and meditation
Guided mental imagery	Is an intervention involving a practitioner generating mental images in the participant

The common factor across mind-body practices, and perhaps the very basis of such practices, involves controlled breathing and/or breath awareness. Breathwork has been long been associated with wide-ranging brain and body responses. These concepts sowed the seeds for Dr. Benson's work on the relaxation response (Benson and Klipper 1975) and are the foundation of the concepts pertinent to mind-body medicine.

Since Benson's research in this area, there has been an exponential growth of studies from NIH and other agencies focused on supporting mind-body practices and their role in brain health (Clarke et al. 2018). The nature of many of these studies demonstrates a growing movement toward integrating mind-body practices into frontline healthcare, with clinical trials focused on the therapeutic value of such practices. The science of mind-body practices is diverse with implications that include studies demonstrating possible positive outcomes for improving symptoms of cognitive decline. Although a degree of caution is warranted as some methodological limitations to such research has been acknowledged, there is an urgent need for more rigorous research to definitively define therapeutic outcomes for mind-body practices. Nevertheless, available research suggests mind body practices appear to be playing a growing, evidence-based role in precision medicine for brain health by acting as personalized, non-pharmacological interventions that target specific neurobiological, cognitive, and psychological pathways. These practices are increasingly used to moderate brain structure and function, particularly in respect to enhancing cognitive performance through stress reduction and related cortisol levels. Ongoing research is suggestive of impact of mind-body practices on neurological structure and function. One such study (Afonso et al. 2017) found CT head scans revealed significantly greater left prefrontal lobe volume in healthy elderly women who practiced yoga for a minimum of 8 years. Other researchers (Asena and Ezgi 2021) have proposed a regulatory network pertaining to breath work addressing the interaction between physiological, affective, and cognitive processes with potential positive impact in prevention. With regards to impact of mind body practice on cognitive function, Basso and colleagues (2019) found 8 weeks of daily meditation (i.e., 13 minutes of guided meditation) resulted in enhanced scores on measures of attention, working memory and recognition memory. There was also evidence for reduced anxiety symptoms with this meditation practice reflected within this study. A study providing systematic review of 11 studies (Gothe et al. 2019) also offers evidence that mind body practices such as yoga may play a role in mitigating age-related and neurodegenerative cognitive declines via possible reduction of age-related brain atrophy. This review of studies found consistently positive relationships between yoga practice and anatomical regions such as the frontal cortex, hippocampus, anterior cingulate cortex and insula. Yoga practitioners with more experience were noted to have greater positive affect on such neuroanatomical regions than non-practitioners, although positive impact was reported in some studies following yoga intervention (i.e., introduction of yoga practice to inexperienced individuals). Overall, although there is need for ongoing exploration of the clinical benefit of mind-body practices in brain health, available research is strongly suggestive of a positive impact of such practices for both brain structure and function particularly pertaining to cognitive symptoms. From my own clinical perspective, it seems wise to not only educate clinicians and healthcare workers, but also patients regarding the role of mind-body connections in health to empower them as advocates for the integration of such practices into their own healthcare experiences.

Although the movement toward better understanding therapeutic value of mind-body practices is encouraging and exciting, critical review of current literature (Van Dam et al. 2017) reveals considerable methodological limitations in current studies. Researchers and clinicians should openly acknowledge such limitations and the possible adverse reactions to limit public misinterpretation of outcome expectations. This is particularly pertinent now with a seemingly growing public mistrust and misinformation pertaining to science and healthcare. Education about the science and healthcare practices of brain health and disease will serve as an inoculation to such misinformation.

Prevention for Functional Cognitive Disorder (FCD)

Preventing functional cognitive disorder (FCD) mimics that of prevention for cognitive decline and neurodegenerative conditions as described above. This prevention for FCD involves managing underlying co-existing mental health issues, such as stress, depression, anxiety, as well as adopting a brain-healthy lifestyle as reflected within the brain health model discussed earlier. As FCD may stem from trauma and/or chronic stress, building resilience and managing overall health is key along with the dementia prevention strategies cited previously.

References

Afonso RF, Balardin JB, Lazar S, Sato JR, Igarashi N, Santaella DF, Lacerda SS, Amaro E Jr, Kozasa EH. Greater cortical thickness in elderly female yoga practitioners-a cross-sectional study. Front Aging Neurosci. 2017;9:201.

Al-Thaqib A, Al-Sultan F, Al-Zahrani A, Al-Kahtani F, Al-Regaiey K, Iqbal M, Bashir S. Brain training games enhance cognitive function in healthy subjects. Med Sci Monit Basic Res. 2018;24:63–9.

Asena B, Ezgi K. Keeping the breath in mind: respiration, neural oscillations, and the free energy principle. Front Neurosci. 2021;15:647579.

Basso J, McHale A, Ende V, Oberlin D, Suzuki W. Brief, daily meditation enhances attention, memory, mood, and emotional regulation in non-experienced meditators. Behav Brain Res. 2019;356:208–20.

Benson H, Klipper MZ. The relaxation response. New York: William Morrow and Company; 1975.

Chapman S, Bond, Fratantoni JM, Robertson IH, D'Esposito M, Ling G, Zientz J, Vernon S, Venza E, Cook LG, Tate A, Spence JS. A novel BrainHealth Index prototype improved by telehealth-delivered training during COVID-19. Front Public Health. 2021;9:641754.

Clarke TC, Barnes PM, Black LI, Stussman BJ, Nahin R. Use of yoga, meditation, and chiropractors among U.S. adults aged 18 and older. NCHS Data Brief, no 325. Hyattsville: National Center for Health Statistics; 2018.

Eugene AR, Masiak J. The neuroprotective aspects of sleep. MEDtube Sci. 2015;3(1):35–40.

Gothe NP, Khan I, Hayes J, Erlenbach E, Damoiseaux JS. Yoga effects on brain health: a systematic review of the current literature. Brain Plast. 2019;5(1):105–22.

Graham-Engeland JE, Sliwinski MJ, Almeida DM, Engeland CG. Psychological stress and cognitive brain health: policies to reduce dementia risk. Policy Insights Behav Brain Sci. 2025;12(1):94–101.

Livingston G, et al. Dementia prevention, intervention, and care: 2020 report of the *Lancet* Commission. Lancet. 2020;396(10248):413–46.

Livingston G, Huntley J, Liu KY, et al. Dementia prevention, intervention, and care: 2024 report of the *Lancet* standing Commission. Lancet. 2024;404(10452):572–628.

Pahlavani HA. Exercise therapy to prevent and treat Alzheimer's disease. Front Aging Neurosci. 2023;15:1243869.

Potashkin JA, Vidyadhara DJ, Hunsberger HC. The impact of lifestyle on brain health. Am J Lifestyle Med. 2025:15598276251411888. https://doi.org/10.1177/15598276251411888.

Samtani S, Mahalingam G, Lam BCP, Lipnicki DM, Lima-Costa MF, Blay SL, Castro-Costa E, Shifu X, Guerchet M, Preux PM, Gbessemehlan A, Skoog I, Najar J, Rydberg Sterner T, Scarmeas N, Kim KW, Riedel-Heller S, Röhr S, Pabst A, Shahar S, Numbers K, Ganguli M, Jacobsen E, Hughes TF, Crowe M, Ng TP, Maddock J, Marseglia A, Mélis R, Szcześniak D, Wiegelmann H, Vernooij-Dassen M, Jeon YH, Sachdev PS, Brodaty H. SHARED consortium for the Cohort Studies of Memory in an International Consortium (COSMIC). Associations between social connections and cognition: a global collaborative individual participant data meta-analysis. Lancet Healthy Longev. 2022;3(11):e740–53.

Seabrook JA, Avan A, O'Connor C, Prapavessis H, Nagamatsu L, Twynstra J, Stranges S, MacDougall A, Hachinski V. Dietary patterns and brain health in middle-aged and older adults: a narrative review. Nutrients. 2025;17(9):1436.

Spira AP, Chen-Edinboro LP, Wu MN, Yaffe K. Impact of sleep on the risk of cognitive decline and dementia. Curr Opin Psychiatry. 2014;27(6):478–83.

Van Dam N, Van Vugt M, Vago D, Schmalzl L, Saron C, Olendzki A, Meissner T, Lazar S, Kerr C, Gorchov J, Fox K, Field B, Britton W, Brefczynski-Lewis J, Meyer D. Mind the hype: a critical evaluation and prescriptive agenda for research on mindfulness and meditation. Perspect Psychol Sci. 2017;13:36.

Compensatory Symptom Management for Behavior and Cognition

11

In my own clinical practice, the review of results in follow-up after conducting a neuropsychological assessment is most often focused upon teaching cognitive compensatory strategies to patients and their families to facilitate their management of cognitive concerns, regardless of cause. Cognitive compensatory strategies refer to the techniques, tools, and environmental changes that may be implemented to "compensate" or work around cognitive weakness, often by using cognitive strengths identified within formalized neuropsychological testing. Generally, the goal of cognitive compensatory strategies is to optimize cognitive function and independence in daily life, rather than to reverse or "fix" the underlying cognitive deficit (Kim et al. 2025). Generally, cognitive compensatory strategies can be grouped into two categories:

1. *Internal Strategies*—These involve making behavioral change in individuals that serve to positively impact cognitive performance on a functional basis.
2. *External Strategies*—These refer to environmental aids (i.e., external to the individual) that are designed to support optimal cognitive performance on a functional basis.

Typically, patients and their famillies may require guidance for use of both internal and external strategies in order to effectively navigate their cognitive concerns. To inspire appropriate use of such strategies, clinicians must engage and support patients, and in some cases their loved ones, to enhance understanding of their own unique cognitive patterns and how they may benefit from implementation of cognitive compensatory strategies. A review of both external and internal strategies to manage cognitive issues is provided below, prior to a more in-depth discussion of cognitive domain-specific strategies that may be implemented for those struggling with cognitive change and/or to empower those who may be suffering from debilitating fear/anxiety for cognitive decline.

S. Hayden, *Fear of Forgetfulness*, https://doi.org/10.1007/978-3-032-21818-6_11

External Strategies (Environmental Changes and Aids) Examples of external strategies for cognitive difficulties include the following:

- *Calendars and Planners*: These include digital or paper day timers, calendars for appointments, and to-do lists.
- *Reminders and Alarms*: These include use of phone alarms, wristwatch alarms, smart speakers for task cues, and in some cases, post-it notes.
- *Organization*: This entails strategies such as designation of specific spots for frequently used items (keys, wallet) and use labels for cupboards.
- *Technology*: It is the use of various smartphone apps and digital reminders (i.e., password apps.).
- *Environment*: This encourages reduction of environmental distractions (i.e., turn off radio when driving) and may involve family in maintaining routines.

Internal Strategies (Metacognitive) Examples of internal strategies for managing cognitive issues are provided below:

- *Self-Talk*: It is the internal dialogue used to guide actions or remember steps (e.g., repeating steps of task or telephone numbers aloud).
- *Visualization*: This entails creating anticipatory mental images of completing a task or remembering an appointment.
- *Information Chunking*: This involves grouping items (i.e., items on a grocery lists) or linking new information to existing knowledge to maximize recall.
- *Management of Attention*: This encourages focusing on one task at a time rather than multitasking. This may also entail limiting the number of tasks requiring focused attention within 1 day.
- *Self-Awareness*: This involves the understanding (metacognition) personal cognitive strengths and weaknesses to better apply the above strategies.

In addition, to both internal and external cognitive compensatory strategies, there may be need for support in managing lifestyle factors such as those discussed within the previous chapters in conjunction with direct training on these compensatory techniques. Training for many of the compensatory interventions directed at managing cognitive concerns may be provided by occupational therapists with expertise in neurocognitive issues. In this regard, occupational therapists can support individuals in task-specific training (i.e., practicing specific tasks using new strategies) as well as social skills training (i.e., using role-play to support communication and social interaction impacted by perceived or functional cognitive change that may lead to social isolation). Typically in clinical settings, occupational therapists may become involved in this capacity following the diagnostic neuropsychological evaluation, where cognitive strengths and weaknesses are identified to inform subsequent intervention.

Cognitive Domain-Specific Compensatory Strategies

Although provision of cognitive compensatory strategies must certainly be personalized to the specific cognitive concerns presenting with each individual, there is benefit to exploring strategies for specific cognitive domains within the context of this book. Certainly, within clinical practice, I will routinely review cognitive strengths and weaknesses along with the associated general cognitive compensatory strategies with patients in the follow-up session of neuropsychological assessments. Patients and their families may not always clearly articulate functional implications of cognitive concerns within clinical settings, and so providing general guidelines for cognitive compensatory strategies in such instances (i.e., of low specificity or insight) may be beneficial.

Regardless, of the etiology for cognitive inefficiencies, providing practical strategies to manage functional issues a patient may be experiencing in daily life may not only serve to optimize functional capacities but may also result in validation of their experience. As mentioned earlier, such validation may be an essential component of managing stigmatization associated with both cognitive and mental health disorders as discussed within prior chapters. This form of strategizing can also serve to empower patients and their families during periods of diagnostic uncertainty, such as those labeled with the prodromal phases of dementia such as MCI and/or those with complex mental health issues impacting cognition such as FCD. Below is a summary of such compensatory strategies contained within tables for each cognitive domain.

Memory and Attention Strategies

Examples of cognitive compensatory strategies for memory concerns are reflected below. These include strategies for managing attention which may directly impact encoding of material for subsequent recall.

Memory/attention
Use reminders and calendars: Setting alarms on electronic devices (i.e., smartphones and computers), keeping a day planner or calendar, or writing notes to log appointments and "to-do" lists.
Establish routines: Consistently placing key items (e.g., keys, wallet) in the same location to reduce the need for recall. Ideally keeping some routine in daily activities (i.e., meals and exercise at same time each day).
Minimize distractions: Working in a quiet environment by closing doors or logging off email to limit interruptions. Driving without conversation or music.
Take notes: Writing down information during conversations to be recalled.
Avoid multitasking: Focusing on one task at a time or limiting the number of tasks to be completed within a day.

Language Strategies

Examples of cognitive compensatory strategies for language issues, both expressive and receptive, are reflected below. Of note, the experts for intervention of language-based issues would be speech-language pathologists, an invaluable resource if available.

Language (expressive and receptive)
Slow down and simplify: Slowing down of communications and simplify language used may be of benefit for some individuals with receptive language issues (i.e., understanding/comprehension). Minimize use of abstract concepts and rely more on concrete language in communications.
Pair verbal with visual modalities of communication: Use gestures, signs, pictures, or objects to support spoken communications for both expressive and receptive language issues.
Wait times: Allow 10–15 seconds after communications to give extra time for processing and comprehension (i.e., for receptive language issues) and/or allow more time for expression of ideas (i.e., for expressive language issues).
Get attention: Ensure the individual is focused on the conversation by engaging eye contact or with light touch if appropriate.

Executive Function Strategies

For executive dysfunction, including issues with problem solving, the following may be considered as possible compensatory strategies. However, given the complexity of executive functions, such strategies require personalization based on specific cognitive concerns.

Executive function and problem solving
Break down tasks: Dividing complex tasks into smaller, manageable steps.
Prioritize: Organizing "to-do" lists to manage time and effort effectively.
Use a problem-solving method: Employing structured approaches like "stop, think, act" or brainstorming solutions to resolve difficult situations.
Organize work around "best time": Scheduling the most challenging tasks for times of the day when energy and focus are highest.
Avoid multiple tasks: Focus on one task at a time rather than attempting to achieve multiple tasks at one time or even in 1 day.
Reduce environmental distraction: Keep environments simple and "clean" both visually and auditorily (i.e., clean clutter; turn off radio when driving).

Visual-Spatial and Processing Speed Strategies

For visual spatial dysfunction and processing speed issues, possible compensatory strategies are provided below.

Visual-spatial/processing speed
Reduce visual clutter: For visual-spatial deficits, ensuring the environment is clean, simple and minimalist may optimize functional performance
Use verbal descriptions: Use verbal descriptions or self-talk to compensate for visual-spatial weakness (i.e., rather than just visually scanning where you parked the car, use self-talk to describe it "I parked near the purple post on the second floor of the parkade.")
Allow extra time: For psychomotor slowing, allow for extra time to complete tasks particularly those that are complex or demanding.
Use assistive technology: Encourage use of assistive technology such as text-to-speech software/tools or audiobooks if reading or visual formats are a challenge.

Overall, compensatory symptom management provides practical, real-world tools that empower individuals to manage cognitive and behavioral challenges and maintain independence in their daily lives. Indeed, researchers have demonstrated higher levels of neuropsychological function along with higher level of independence were associated with more frequent use of compensation strategies in older adults (Tomaszewski Farias et al. 2018).

Use of cognitive compensatory strategies may also contribute to higher resilience and greater feelings of empowerment in the face of cognitive decline. For patients I serve in clinical practice, whether it be those with MCI, an emerging dementia or mental health factors such as FCD, the feeling of empowerment is critical to shift their experience of cognitive "loss" to more controllable aspects of themselves and their lives. For many, the feeling of looming loss and/or mortality can raise their risk for mental health disorders and reduce their quality of life and feelings of self-worth (Maki and Hattori 2019). Thus, engaging patients and their families in cognitive compensatory strategies may result in secondary gains to brain health through improvement in the primary or secondary mental health issues related to cognitive concerns. Discussion of cognitive compensatory strategies is a means of providing clinical support for self-determination in patients (and their families) when they are faced with fear of cognitive and/or functional loss.

For clinicians, recognizing brain health risks in patient populations and mitigating such risks with empowering compensatory strategies serves to ensure provision of quality care. This appears to be important not just for cognitive concerns relative to prodromal or early-stage dementias but have also been found to have value within mid to later stage illness using techniques of cognitive rehabilitation (Kudlicka et al. 2023). Although such interventions do not, and are not intended to, reverse cognitive impairment, these techniques can support those living with dementia to sustain their functional independence for as long as possible.

Cognitive Compensatory Strategies for Functional Cognitive Disorders

Cognitive compensatory strategies for functional cognitive disorders, as for dementia and its prodromes, involve use of a multifactorial approach to symptom mitigation including but not exclusive to the following:

- *External Aids*—Encourage use of external reminders or aids such as use of calendars, lists, apps, and alarms. Organizing the environment (i.e., designated spots for keys) is beneficial. This may minimize reactivity and catastrophic response to perceived or functional cognitive errors often present in FCD.
- *Internal Techniques*—These include interventions such as positive self-talk and visualizations, aimed at enhancing self-awareness in management of daily challenges in attention, memory, and executive function, focusing on practical skill-building to empower FCD patients sense of control.
- *Environmental Adjustments and Lifestyle Changes*—Incorporating positive lifestyle habits such as regulating sleep and exercise is also of benefit to give FCD patients a framework of empowerment to redirect negative self-talk and/or hypervigilance to cognitive decline and fear for forgetfulness.
- *Metacognitive Training*—This facilitates awareness of tendency for self-monitoring and hypervigilance of cognitive challenges in FCD to boost independence. Metacognitive training is emerging as a one of the more promising approaches to treatment for FCD patients (Bhome et al. 2019), particularly as metacognitive deficits are a prominent feature of FCD. The target of metacognitive interventions in FCD is primarily the misperceptions regarding cognitive abilities (i.e., perceived cognitive deficits despite intact cognitive performance on neuropsychological testing), rather than their own assumption of structural neurological deficits (i.e., assumption that they have dementia). The aims are to improve factors such as excessive self-monitoring, attention regulation, and memory appraisal in order to reduce distress and subjective cognitive complaints (Mavroudis et al. 2025). Metacognitive training is often used in conjunction with Cognitive Behavioral Therapy (CBT) which focuses on issues pertaining to health anxiety and any associated avoidant behaviors. The metacognitive training techniques for FCD include the following:
  - *Psychoeducation*—This helps to normalize cognitive variability that is present for most people as well as explaining the function, non-degenerative, nature of symptoms related to underlying health anxiety.
  - *Self-monitoring/feedback*—This involves training individuals with FCD to objectively evaluate their cognitive performance rather than relying on emotional appraisals of errors (i.e., "I forgot my keys so I must have dementia"). Psychoeducation and review of neuropsychological test findings, if available, may be particularly beneficial in training appropriate self-monitoring techniques for FCD patients.

- *Attention regulation*—This provides FCD patients with skills to shift focus away from internal work and onto external tasks (i.e., positive use of distraction techniques). This may include encouraging engagement in brain healthy hobbies and activities such as mind-body practices discussed in prior chapters.
- *Computerized training*—This introduces software and technology to provide FCD patients with feedback on cognitive tasks in order correct their biases and misperceptions regarding memory and/or cognitive deficits.

Such techniques may be incorporated into structured modules of treatment for FCD patients, although resources in this area remain limited. Further research into efficacy within FCD patients presenting to memory clinics is certainly warranted.

Final Thoughts on Fear of Forgetfulness

My hope in the writing of this book is to provide a broad overview of the complex factors that contribute to the "fear of forgetfulness" experienced by many individuals within the context of advancing age. Perhaps this may inspire such individuals to further explore topics discussed here in greater depth in order to enhance their own brain and mental health experiences. With this intention of further exploration, I have provided just a few of the available readings and resources for this purpose at the end of this book. As has been repeatedly stated throughout the book, psychoeducation of patients and their loved ones on all the confounding factors inherent to diagnostics of neurodegenerative disorders may be the basis to explore more effective means to their management of mental health issues underlying conditions such as functional cognitive disorder (FCD). In many instances, such education and clinical communications are best facilitated by comprehensive, considered neuropsychological assessment, including thorough assessment of both cognitive and mental health symptoms in those with cognitive concerns presenting to memory or dementia clinics.

Further, healthcare providers working within clinics evaluating dementia and their prodromes must now fully acknowledge the relatively high prevalence of mental health conditions such as FCD in such clinical settings. Despite challenges related to the identification of FCD in memory clinics, clinical care can aspire to explore more effective diagnosis and management of such patients. Increased understanding of mental health factors such as FCD, in those presenting for dementia investigations, may facilitate reduction of associated stigmatization of such disorders, including within these healthcare settings. Clinicians working within healthcare settings, such as memory clinics, must strive to inform their own clinical practices through evidenced-based understanding of all confounding factors in order to minimize potential for iatrogenic harm while investigating neurodegenerative disorders.

References

Bhome R, McWilliams A, Huntley JD, Fleming SM, Howard RJ. Metacognition in functional cognitive disorder – a potential mechanism and treatment target. Cogn Neuropsychiatry. 2019;24(5):311–21.

Kim J, Murphy S, Abou L, Alschuler K, Ehde D, Fritz N, Kratz A. Understanding use and predictors of compensatory cognitive strategies in people with multiple sclerosis. Mult Scler Relat Disord. 2025;102:106710.

Kudlicka A, Martyr A, Bahar-Fuchs A, Sabates J, Woods B, Clare L. Cognitive rehabilitation for people with mild to moderate dementia. Cochrane Database Syst Rev. 2023;6(6):CD013388.

Maki Y, Hattori H. Rehabilitative support for persons with dementia and their families to acquire self-management attitude and improve social cognition and sense of cognitive empathy. Geriatrics (Basel). 2019;4(1):26.

Mavroudis I, Petridis F, Franekova K, Visternicu M, Rarinca V, Burlui V, Ciobica A, Novac B, Dobrin I, Hogas MM, Bovari E, Albert C, Kazis D. Functional Cognitive Disorder: diagnostic challenges, clinical features, and future directions in a misunderstood condition. Front Aging Neurosci. 2025;17:1665510.

Tomaszewski Farias S, Schmitter-Edgecombe M, Weakley A, Harvey D, Denny KG, Barba C, Gravano JT, Giovannetti T, Willis S. Compensation strategies in older adults: association with cognition and everyday function. Am J Alzheimers Dis Other Dement. 2018;33(3):184–91.

Recommended Resources

Below is a list of recommendations to further learning and understanding of cognitive and mental health issues.

Books

Bluming A, Tavris C. Estrogen matters. 1st ed. Little, Brown Spark; 2018.
Goleman D, Davidson RJ. Altered traits: science reveals how meditation changes your mind, brain, and body. Illustrated ed. Avery; 2017.
Jordan, M. Coping with mild cognitive impairment (MCI): a guide to managing memory loss, effective brain training and reducing the risk of dementia. Jessica Kinglsey Publishers, London (2020).
Mosconi L. The XX brain. New York: Avery: Penguin Random House; 2020.
Rowe ET. Functional neurological disorder workbook: a FND mind-body recovery. Therapybooks Publishing; 2025.
Sapolsky R. Why zebras don't get ulcers. New York: St. Martin's Griffin; 2004.

Websites

Alzheimer's Disease International. https://www.alzint.org/reports-resources/.
American Association of Retired Person (AARP)/Global Council on Brain Health. https://www.aarp.org/health/brain-health/global-council-on-brain-health.
American Psychological Association. https://www.apa.org/gradpsych/2009/03/cover-abroad.
Calm app. https://www.calm.com.
FND Hope (Functional Neurological Disorders). https://fndhope.org/fnd-guide/.
HeadSpace app. https://www.headspace.com.
Mementum app. https://www.cogniss.com/cogniss-app-showcase/mementum.
Mindfulness apps. https://mindfulnessapps.com.
Women's Brain Health Initiative. https://womensbrainhealth.org/.

S. Hayden, *Fear of Forgetfulness*, https://doi.org/10.1007/978-3-032-21818-6

Index

S. Hayden, *Fear of Forgetfulness*, https://doi.org/10.1007/978-3-032-21818-6

MIX
Papier aus verantwortungsvollen Quellen
Paper from responsible sources
FSC® C105338

If you have any concerns about our products,
you can contact us on
ProductSafety@springernature.com

In case Publisher is established outside the EU,
the EU authorized representative is:
Springer Nature Customer Service Center GmbH
Europaplatz 3, 69115 Heidelberg, Germany

Printed by Libri Plureos GmbH
in Hamburg, Germany